Raising Goats

Beginners Guide to Raising Healthy and Happy Goats

By: Janet Wilson

Table of Contents

Part I: Introduction and FAQ

Introduction

Goats are wonderful animals to raise. Compared to other livestock animals, they are easy to care for, not expensive, don't require much water, and are natural-born comedians. Every time I walk into my goat paddock, I laugh.

Humanity and goats have an ancient partnership. Archeologists have found evidence that goats have been living with humans as far back as 6000-7000 BCE. I have found that knowledge of this long history encourages me to reflect on our ancestors and makes my goat raising even more rewarding.

We will start with some FAQs, move to what you need to set up and get started with your goats. After that, we will discuss maintenance and breeding.

There are more than 200 breeds of goats; this book is a curated list of breeds by their *function.* Do you want goats who are good pets? For milk? For fiber? We will discuss those functions and more and recommend common breeds that are highlighted for their traits.

Frequently Asked Questions

Can I have goats?

You need to check with your local municipality to determine whether you can have goats at all and, if so, whether there are restrictions. In the US, goats are listed as "livestock" and, as such, can only be kept in premises zoned for agriculture. I am not aware of any exceptions to this, but laws change all the time.

If you live in a suburban area with large backyards that could accommodate goats, find out if they are allowed. If they are, make sure to find out about regulations and restrictions. You might be able to have goats, but the number might be limited, as may the size. Sometimes residential areas with a few acres of space may border on agricultural zones but be zoned as residential and not allow goats.

The backyard homestead movement has grown increasingly in numbers and strength over the last 15 years. There may have been people who have already advocated for changes in zoning laws in your area.

Ask yourself: Are there dogs who may harass or be a danger to my goats?

Sadly, domestic dogs are the most common predators and cause the highest numbers of predatory goat deaths.

If you have a dog who likes to chase, it can exhaust a panicked goat for fun even if it is a small dog. If you have a dog and have not yet had goats, then make sure that you can keep the dog out of the paddock and separated from the goats except under supervision. In the dog section, we discuss training for dogs.

If you don't have a dog yourself but live in an area where there are dogs who run free, it is crucial to protect your goats. If you live where dogs might get out in your neighborhood, you might consider a 6-foot instead of a 4-foot fence. Sometimes neighborhood dogs in quiet areas will be buddies and keep each other company. This is lovely until their brains flip into a primal chase/predator mode and work together as a pack to attack your goats.

Assess the risk, build fencing accordingly, train your dog, and make sure you can keep them away from the goats if necessary. We discuss dogs more in the FAQs below as well as in Part Three.

How many goats should I get? Can I get just one?

Goats are herd animals and will not be happy without at least one other goat. They don't have to be the same breed of goats. When we started to build out our last herd, the first goat I acquired was a Nubian. She was very clingy to me when we got her; we were waiting for other goats in the herd to be delivered. She was also very loud. When the five Nigerian Dwarfs arrived, she was perfectly happy. Get at least two goats. If you are getting pack goats, get at least two of the similar breeds. Even if they have other goats to live with, they are much happier on the trail with another goat.

How much space do I need for goats?

The rule of thumb is that you need 250 square feet of outdoor space per goat. You only need 20 square feet of space per goat for sleeping and shelter because they like to be close when sleeping and resting.

If you happen to have land with a large pasture or wooded/brushy area to roam during the day, your goats may only need 15 square feet of space per goat in their shelter. This calculation depends on that outdoor space being available all year round. If you live in a place with very cold or wet winters where the goats may spend a lot of time in the shelter and not be out browsing so much outside, it is best to give them 20 square feet per goat in their shelter.

How much time and work does it take to raise goats?

That depends on what you are raising your goats to do.

If you have a couple of goats as pets, then you'll need about 5-10 minutes a day to change their water, give them some feed, handle them to check that they are OK. That's the minimum. If you want your goats to be pets, though, you'll want to handle them a lot, so they are docile and tame. You'll want to spend time with them so that you have a close bond, and, of course, because they are affectionate, fun, and hilarious.

If you have dairy goats that you are milking twice a day, you'll need more time. If you have 3 or 4 dairy goats, you will need about 1 hour twice a day for all of your chores.

If you have brush goats who you rotate around your land, then you'll need to do all the water filling and health checking/handling, plus leading them out to their browsing area for the day. You will probably need to set-up the fencing also. If you choose to use an electric fence, then setting up and maintaining the fence system will also be part of the work.
Other goat functions vary in their time requirements. We will discuss this as we go through the various breeds. However, for all goats, every 6-8 weeks, you'll need to trim their hooves or take them to the vet for trimming. That takes about 15 minutes per goat.

Do goats smell?

Mostly, no. Does and wethers (castrated males) don't smell any more than any other animal. A goat's smell is only about as strong as an average dog (even less), just a different smell; it's not unpleasant. Some people let miniature goats inside their house.

The exception to this is a *big* exception: Bucks, aka Billy Goats (uncastrated males) smell. Actually, they *stink*. Bad. Bucks have a hormonal musk that is attractive to does but repellent and disgusting to humans. When they are in rut, they add to the musk odor by peeing on their faces, beards, and chests.

Unless you have a buck, you won't have trouble with your goats being smelly if their environment is kept clean, and they get wiped off with a damp towel once in a while when necessary.

Can people get sick from goats?

Yes, you can get sick from goats; *however*, you can prevent it with excellent basic hygiene practices. Here are what I call my "Basic Barnyard Protocols". (I have never gotten sick from goats, and I don't know anyone who has.)

Basic Barnyard Protocols

These Basic Barnyard Protocols are measures for common sense prevention. Some diseases are transmitted through direct contact; others are airborne, or both. There are essential best practices for any barnyard that will virtually eliminate your risk.

- Wash your hands thoroughly after petting, handling goats **and** after going through gates or handling equipment. Teach your children to do the same, supervise children's handwashing if they are under five.

- Have a separate pair of boots and gloves that do not go inside to wear for the chores in your goat paddock.

- There may be times when it makes sense to wash your clothes after handling goats and cleaning out their paddock. The risk for transmission is obviously higher if you know a goat is sick, if you just acquired a new goat and are unsure of its health background, or if you needed to kneel or sit on the ground where they pee or had to handle urine-soaked straw. Assess the risk and use common sense.

Diseases that can Transmit from Goats to Humans

Chlamydiosis

Chlamydiosis is a bacterial disease that is sexually transmitted between goats. When an infected doe gives birth or aborts, the liquid and tissues (e.g., placenta) are particularly high in this bacterial content. It is not common for humans to get it, but can happen if precautions are not taken when in direct contact with birth/miscarriage tissues and fluids.[1]

Goat Symptoms
The main symptom of chlamydiosis is miscarriage or stillbirth. Goats might carry chlamydiosis and never have any symptoms. Other symptoms may include conjunctivitis (pink eye), respiratory conditions, or arthritis.[i]

Human Symptoms
Women may notice unusual vaginal discharge or a burning sensation when urinating. This is easily confused with a urinary tract infection; it is critical to diagnose accurately. Men may also feel burning when urinating or have a discharge from the penis.

Prevention of chlamydiosis transmission
1. When kidding (birthing goats), use gloves, don't touch your eyes or face while out with the goats, and avoid splashes. Of course, wash your hands after being out. Disinfect the area and your boots after kidding or miscarriage.
2. Keep other animals away till the area has been cleaned and disinfected.
3. Pregnant women must ensure that they are not exposed. It is recommended that they not handle pregnant or aborting does.[2]
4. If you want to breed your doe(s) with a buck outside your herd, determine whether the herd he comes from has had miscarriages or stillbirths. If he has been lent out to sire in other herds, most goat raisers and vets will recommend that your does do not breed with this goat.

Q Fever

Q Fever is a bacterial infection. Its transmission is usually airborne, but direct contact can sometimes transmit it as well.

It is carried mostly by sheep, goats, and cattle. The bacteria are most highly concentrated in birth tissues and fluids, but they are also carried in manure, urine, milk, and blood. It is known to be very contagious, and it doesn't take much concentration in the air to infect a human just through the dust in the barnyard.[3] Direct contact with any of these substances or drinking raw, unpasteurized goat milk, can transmit it. If you are immune-compromised, are pregnant, or have a heart condition, you are more susceptible to Q Fever.

Goat Symptoms
The animals are often not symptomatic. The main sign you will notice is miscarriage. Q Fever is one possible cause.

Human Symptoms
2-3 weeks after exposure to the bacteria, a person feels flu-like symptoms. 1 in 20 people who get Q fever develop a chronic version that infects the heart valves. See your doctor if you expect you have contracted Q fever.[4]

Leptospirosis

Leptospirosis is a bacterial infection. You can inhale leptospirosis when cleaning the paddock. It is also transmitted through direct contact with the wounds or mucous membranes of infected animals. Soil, water, urine, feces, soiled bedding can all carry leptospirosis bacteria.

Goat Symptoms
A goat might signal leptospirosis with anemia. Besides that, the sign will be a miscarriage.

Human Symptoms
The CDC notes that humans will have sudden flu-like symptoms 2 days-4 weeks after contracting the disease. Also, you may think you have the flu and go through a "phase 1", then become ill again. It is crucial to determine whether it was leptospirosis as "phase 2" can lead to kidney or liver failure and even meningitis.[5]

Prevention
The most common way humans get leptospirosis is through contact with water, soil, goat bedding, urine, feces, or other infected animals' body fluids. Your basic barnyard protocols are your strongest prevention. [6]

Ringworm

Ringworm is a fungal infection. You can get it from direct contact with the goat and equipment such as hoof trimmers, halters, saddles, or blankets. You can also spread it through your clothing.[7]

Goat Symptoms
It is called "ringworm" because circular, crusty lesions will appear. The fungus attacks the skin and hair/fur. The animal may clue you to it because they are very itchy.

Human Symptoms
Same as the goats. You will have an itchy, crusty circle on your skin.

Prevention
Clean and sanitize all equipment that comes into physical contact with your goats. This includes your clippers. Also, wash or sanitize blankets, saddles, halters and anything that is close to the animal's skin to avoid infection.

Salmonella

Salmonella is a bacterial infection that lives in the intestine. Salmonella is most famous for being transmitted through eating contaminated food that has not been thoroughly cooked or is unpasteurized. It can be passed on through the fur and direct contact with feces or bedding or soil with feces on it.

In 2018, France reported that 150 people were affected by a Salmonella outbreak from raw goat's milk cheese.

It is not common for goats to pass on salmonella, much more common to get it from chickens. However, if your goats share a paddock with chickens, this raises the risk considerably and makes it all the more dangerous to consume raw milk or cheese.

Goat Symptoms
Because salmonella affects the intestine, the symptoms are more obvious than the other transmittable diseases we've discussed.

The first sign will probably be diarrhea. Always get a stool sample to your vet if one of your herd has diarrhea and isolate them until the cause is known.

The symptoms can quickly go to fever, lethargy, perhaps being "tucked-up and miserable". Other symptoms are kicking at their belly or grinding their teeth.[8]

Human Symptoms
Humans have diarrhea, fever, and stomach cramps. The incubation time can be as little as 6 hours to as long as 6 days or even several weeks. Like the goat, we may want to tuck up, kick our own belly and grind our teeth.[9]

More severe cases can last longer (months rather than weeks) and lead to "reactive arthritis".

Salmonella comes from either eating or touching infected materials or animals. Prevention relies on basic barnyard protocols such as washing hands, cleaning equipment, pasteurizing milk and cheese, and fully cooking meat.

Practice your own best barnyard protocols for cleanliness. Salmonella can live for months to years in an environment that is moist and left unclean.

Are goats friendly to humans?

Goats are not only friendly but delightful. With the right handling, many will bond to their humans, similar to a dog. As with dogs, breeds and individuals vary. Overall, goats are an easy barnyard animal to have. Through this guide, you'll find out more about how to enjoy their friendly nature, intelligence, and silly antics.

Do goats get along with dogs and other species?

Yes! Goats are very social and need at least one other goat, but they also get along with other barnyard animals such as sheep, cows, horses, donkeys, and chickens. They like to be social with everybody. They are fine with cats, and if a dog is friendly to them, they will enfold the dog into the herd as a friend and protector. The dog might find the goats snuggling up to them.

Being prey animals, goats are cautious. Individual goats who have been harassed or attacked by a dog before might reject interaction with any dog and exhibit signs of stress in their presence. If you have a dog, this is useful information to have about any goat that you acquire.

There are breeds of dogs that are more or less suited to being with goats, and, of course, individuals will vary as well.

If your dog sees the goats as prey, the dog needs training or reliable separation from the goats. You may be able to train the dog, but never leave them in the presence of the goats alone.

We will discuss more details about introducing dogs and goats in "Part Two: Getting Started".

Do all goats have horns?

Yes. Goats have horns. Well, most goats have horns. They are prey animals, and their horns are their protection. The horns also regulate their temperature. They are filled with blood vessels. Therefore, in hot climates and activities that require a lot of exertion, horns are a big plus. If you don't want your goats to have horns, a vet or reputable breeder will be able to remove the horns from very young goats. You can also learn to do it yourself. Disbudding (horn removal) is discussed in the breeding and birthing section.

Goat vocabulary

Here are some terms you will read throughout this book:
- **Caprine** – This is the technical term for goat. Equivalent to "bovine" for cow or "ungulate" for deer. It comes from the Latin, *caprinus*.
- **Cleat** – Goats have two parts to their hooves, called cleats.
- **Wattle** – Extra skin that hangs from the necks of some breeds of goats.
- **Withers** – The point between the shoulder blades. The height of the goat is measured at the withers.
- **Polled Goat** – A goat born without horns.
- **Disbud** – To disbud a goat is to burn off their horn's buds when they are very young.
- **Paddock** – The enclosure where a herd of goats is kept.
- **Buck** – A male goat who is of breeding age.
- **Buck Power** – The number of does a buck can breed in a season.
- **Billy Goat** – A buck.
- **Buckling** – A young male goat who is not yet ready to breed.
- **Wether** – A castrated male goat.
- **Doe** – A female goat that has not yet given birth.
- **Doeling** – A young doe who is not yet ready to breed
- **Dam** – A female goat who has given birth.
- **Nanny Goat** – A dam.
- **Freshening** – When a female goat gives birth, it "freshens" her milk production.
- **First Freshener** – The first time a doe gives birth.
- **Kids** – Baby goats
- **Throwing** – Giving birth. "Our doe threw two kids yesterday."

- **Kidding** – The process of labor and giving birth
- **Marking Harness** – Used for breeding. An apron on the chest and stomach of a buck that is designed with a marker. Each buck in the herd has his own color and when he mates with a doe, she is marked so the owner can track the lineage in their herd.
- **Estrus or "heat"** - 12 to 36 hours in the estrus cycle when the doe will stand and allow the buck to breed her. [10]
- **Estrus Cycle** - The period from one heat cycle to the next. The estrous cycle of goats occurs every 18 to 24 days, or 21 days on average. Some breeds are seasonal, others cycle all year round. [11]
- **Colostrum** – The thick fluid produced by the mother goat at birth before milk production comes through.
- **Raw Fiber** – Raw fiber is the material combed out of the goat with no processing whatsoever.
- **Processed Fiber** – Processed fiber has been dehaired, washed, and carded.
- **Virgin Fiber** – Fiber that has been made into yarn or material for the first time.
- **Recycled Fiber** – Fiber that has been reused/repurposed from another fabric or yarn.
- **Guard Hairs** – The longer, stiff hairs that grow through the undercoat of a goat.
- **Type A Fiber** – Mohair from Angora goats only
- **Type B Fiber** – Cashgora – Fiber from goats who are mixed breeds between Angoras and a cashmere goat. Popular cashgora goats are Nygoras (Angora/Nigerian Dwarf mix) and Pygoras (Angora/Pygmy mix)
- **Type C Fiber** – Cashmere, the undercoat of any goat that grows an undercoat.

Part 2: Getting Started

The first thing to do to get started raising goats is to decide *why* you want goats! There are a wide variety of reasons that people raise goats. Do you want a couple of miniature goats for family pets? Do you want to milk goats and make cheese? These are only a couple of goat functions, and there are breeds designed to fulfill those functions. Here is the list of purposes and functions to consider when choosing breeds:

- Pet Goats
- Brush Goats
- Dairy Goats
- Fiber Goats
- Working Goats
- Show Goats
- Meat Goats

Depending on how many goats you have, you can have multiple functions in your herd as well. We use our goats for brush-clearing, pulling carts, carrying packs, and one for milk. The brush around our land is at a varied height, so we have Nigerian Dwarfs who get the low and medium stuff, and the higher brush is gobbled up by the working goats and the Nubian. If we wanted to make an effort, we *could* also use these goats as fiber goats. The Nigerian Dwarfs are very fun pets to my two young children, and I am closely bonded to the Nubian, similar to a dog.

There is a lot of great information available on the internet and in books about raising goats; you can absolutely learn yourself and get started. However, *if you depend on your own research, the key to success is to start small.* We highly suggest getting your equipment and systems set up, getting two goats to start with, then building your herd from there.

You may have an experienced local mentor who can be readily available to you as you get used to keeping and tending your new goats. If that is the case, then ask them how many goats they would recommend you start with.

After you decide what breeds you want, then you can turn your attention to preparing the space for them.

A word about Billy Goats (aka bucks)

- During their rut, bucks have a musk to attract females that is repulsive to humans. People use words like "vile", "stench," and "disgusting" to describe it.
- Not only do they have a musk, but they also pee on their faces, beards, chests, and legs to attract does. Then they jump up on your fence or rub on you.
- You'll need a place where you can separate the buck(s) from the does so they don't harass and stress the does and breed too much.[12] Even outside of the rutting season, if you use does as milking goats keep them separate because the buck will rub all over them, and you'll have to smell it while you're milking. It can also come through in the flavor of the milk. Not Appetizing.
- The separate space is also good for timing the births in your herd. You don't want your does to be kidding all at the same time, nor do you want them breeding too early or too often. (We will go into more detail about this in the breeding section.)
- If you decide that one year you don't want to breed your does, the buck will bellow when he goes into rut. He will be very noisy, obsessed, and even go off his food.
- Breeds and individuals vary, but bucks have a tendency to be aggressive to both humans and the herd.
- If you decide to have a buck, also have at least one other buck or wether who can hold his own against the buck when the buck goes into rut. When you separate the buck, put the other buck or wether with him so that he's not alone. *Goats need at least one other goat with them.* If the buck is alone, it will only bring out the worst in his temperament.
- Note that "separate space" does not mean on the other side of the fence. Bucks are determined to breed, and when a female is in heat, they will breed through a fence.

When a buck is castrated and becomes a wether, his temperament turns into that more like a doe. He becomes cooperative, tame, calm, does not stink, fits right in with the does, and will not harass them. "Wethers are Wonderful" would make a great bumper sticker.

Keeping a buck is a choice you may want to make in order to breed your goats, but as you can see, they can be a "handful". They also require extra space and consideration. We suggest starting with a couple of does or wethers and then add a buck when it's time for breeding so that they keep producing milk.

As we discuss each function of goats, we will list breeds that are the best for that function. Included are notes about any attributes or tendencies of the bucks that are specific to that breed, but for a beginner, we recommend starting with does and wethers.

Functions of Goats and Best Breeds

Out of the more than 200 domestic goat breeds in the world today, most don't fit neatly into function categories – they tend to be multi-functional. The lists below are according to the *best of the common breeds for each function.*[13] In each section, we discuss the skills and knowledge required for the function (such as how to manage dairy goats, how to use your brush goats for clearing, how to harvest the cashmere from fiber goats, etc.).

If you decide you want your pet goats to also give you milk, make sure that you seek out a breeder who is breeding for milk production as well as temperament for pets. If you get one that is only breeding for temperament, you may be disappointed with the goat's milk production.

Let's start with the easiest goat function; you may just want to have some pet goats.

Pet Goats

Pet goats are incredibly fun for anyone and a wonderful way to introduce a young child to the responsibilities of animal care. Some parents who have larger livestock get miniature pet goats for their young children to introduce them to caring for livestock.

The care and skill level required for pet goats is low.
- They need handling and interaction every day so that they bond with you.
- Consistent training is required so that they don't butt other goats or humans and don't jump up on people. Like a dog, clarity, and consistency while being firm yet kind produces the best results. (See the training section in maintenance for more detail)
- You'll need 5-10 minutes a day to make sure they have plenty of clean, fresh water at all times and to feed them.
- When you handle them, give them a simple health check. (See the maintenance section for more details)
- Smaller goats require fencing that does not allow them to wiggle through; glance at your fencing each day to make sure no clever goat or predator did any damage.

One small way that they require more attention and maintenance is hoof trimming. Pet goats are usually not getting the wearing on their hooves from rocks and long distances in brush. This means that if your breed says "needs trimming every 4-6 weeks," it is important to check at 4 weeks because it is likely your goats will be on the short end of the interval rather than the longer end.

Your pet goats will need an outdoor space. You *might* be able to train your goats to use a goat litter box for peeing (see the training section), but you will *not* housetrain them away from pooping inside. Also, remember goats chew on things, and a number of breeds love to jump up on surfaces to get high up to see. You're likely to have goats on your tables, piano, couches, and chairs. At the very least, you'll have chewed up furniture (chewed up everything, actually) that has been jumped and pooed on.

If you have the space, I've known people who made a sheltered outdoor lounge with old furniture so they could have a "living room" space to comfortably hang out with their goats. They even watch movies and TV on their laptops. Reading is hard because goats love paper, and they will want to eat your book and tear out the pages.

If you just want pet goats and are not interested in breeding or milking, then whatever breed you choose, we suggest getting wethers. Bucks smell and are obsessed with breeding. Does go into heat (some breeds every month) and can make a lot of very loud noise when they want to mate. Pet wether goats will release you from both of these issues and be lovely companions.

What kinds of goats make the best pets? Here are eight favorites.

Pygmy Goat

Wide Open Pets http://ow.ly/bIaf50CH9fB   Ecstasy Coffee. http://ow.ly/gMlw50CH9vJ

General Description

Pygmy goats originated and were bred in West Africa for milk, meat, and docile temperaments. After being introduced in the US in the 1950s, the combination of their "cuteness factor" and personalities made them popular as pets. The Pygmy build is a bit stockier than their relatives, the Nigerian Dwarfs.

Height: 15-20"
Weight: Females: 35-50 lbs Males 40-60 lbs.
Lifespan: 12-18 years, 15 years is average.
Cost: Spring babies are $75-$500. You pay top prices for registered animals who are show worthy.

Temperament

Friendly, social, affectionate, docile, sweet, tame, gregarious, hilarious, a constant source of entertainment and joy.

They love to dance, prance, and seem to be born to hop sideways - they do this when they are just a couple of days old. Modern Farmer put together a few videos of Pygmy prancing. Enjoy this.[14]

The mix of their temperament and size makes them especially good pets for small children.

Climates

Adaptable to all climates, but they do **not** like being wet. Especially take care to keep their hooves dry and out of the mud.

Prone to any illnesses or conditions?

Prone to bacterial hoof conditions such as "hoof scald" and "hoof rot". Hooves need to be kept dry, clean, and trimmed. Care needs to be taken that bacteria does not get trapped in the hoof.

Special Considerations

- As mentioned earlier, Pygmy goats hate getting wet. They originated in the arid climate of West Africa. Do not try to wash or soak your goats; wipe them off, and "spot clean" if they need it.

- They do not like sleeping on the ground. Give them a raised wood platform to sleep. This also keeps their hooves out of moisture all night and prevents acute hoof disorder.

- When it's muddy or wet snow, make elevated places where they can walk about in their outdoor enclosure. Use planks of wood, bricks, anything that will give them a way to walk out of the mud or wet. They don't like wet in general but *really* hate to get their hooves wet or muddy.

- Their fencing needs to be 4' high and also designed so that they can't slip through small spaces. A cow panel fence has spaces that are too big.

- They are very active and spend most of their time playing. Provide stimulation, various heights of platforms, or tree stumps so that they don't get bored.

- They breed all year. If you don't want to be handling constant reproduction, make sure you get does or wethers. The does go into heat regularly and will cry and bleat while they want to mate.

- They are particular about their water. Their water must be kept clean and fresh. If their water is dirty, they won't drink at all.[15] This is true of all goats but is intensified in Pygmys.

- All goats love to be up high, but Pygmys especially love to get on top of high places to see. Provide some wooden platforms for them so they can get a view. Even though they aren't "jumpers", keep those platforms well away from the fence.

Feed

All goats benefit from more vitamin A & D in the winter, but for some reason this is especially true for Pygmys[16]

Health

Talk to your vet about annual vaccines and deworming if needed.
Trim the hooves every 4-6 weeks. Pygmys are prone to hoof rot more than other breeds.

Multi-functions

Despite their size, farmers actually love Pygmy goats for milk production. A doe can give you 2-3 quarts of milk a day![17] The butterfat is one of the highest in goat breeds (6-11%) so it is great, high-quality milk.

Pygmy goats can also be meat goats.

Pygmy goats are great to have as part of a brush clearing goat brigade. They will work on the lowest part of the brush that larger goats may ignore.

Beginner Friendly

Yes! In fact, Pygmys are recommended as one of the best (or the best) breed for your first goats.

Pygmy Resources

National Pygmy Goat Association
https://www.npga-pygmy.com/default.asp
There are lots of good resources here and reportedly the people in this association are very helpful to beginners.

Registered Pygmy Goat Breeders in the US
https://www.npga-pygmy.com/contacts/breeders.asp

Facebook Rocky Mountain Pygmy Goat Club
https://www.facebook.com/groups/946364028748574/

Willamette Pygmy Goat Club
http://www.wpgc-pygmy.com/

Gold Country Pygmy Goat Club
https://goldcountrypygmygoatclub.com/

Nigerian Dwarf Goats

Farming My Backyard http://ow.ly/DcqC50CH9Jt
Dreamers Farm http://ow.ly/O0MW50CH9Ok

General Description

Nigerian Dwarfs are a little bigger than the Pygmys, and even though they are a little "round", they are slenderer in build. Nigerian Dwarfs also came from West Africa, but they had a unique journey to make their way into the hearts of American goat raisers.

The Nigerian Dwarfs were bred in Sudan as long as 5500 years ago. They were brought from Africa on boats to be food for big cats who were being transported to the US for Zoo exhibitions. The goats who were not cat food were given to the last zoo on the drop-off stops. It took some years to distinguish between the stockier Pygmy Goats and the Nigerian Dwarfs who were more like dairy goats, but gradually the breeds were recognized and separately registered. The Gladys-Porter Zoo in Texas received many of these goats throughout the years.

If you are looking for a pedigree whose ancestry can be traced back to the Nigerian Dwarfs who came to Texas, then Sharla Parker and Kathleen Clapp as well as the Gladys Porter Zoo will be noted in their papers. [18]

Height: Does 16"-21" Bucks 18"-23"
Weight: 60-80 lbs.
Lifespan: 12-14 years when well cared for.
Cost: $50-$500. Those considered show quality will be much more. "Pet quality" goats are on the lower end of the price range.

Temperament

Nigerian Dwarfs are gentle and docile as well as good with their herd and other barnyard animals. They are wonderful for children, and fun members of any family. They are even known to be good companions to aging livestock such as horses. [19]

It is a joy to get to know each individual Nigerian Dwarf as they present with different personalities, likes and dislikes.

As with all goats, Nigerian Dwarfs are intelligent, and it is important to interact with them a lot and give them toys and things to do so they don't get bored. They need a way to explore and follow their curiosity. They are also very energetic and active, so they need exercise and physical challenges like things to jump up on, tunnels, or agility training.

Climates

Like the Pygmys, they adapt to any environment. I currently have seven goats of this wonderful breed, the temperatures here range from below zero to over 100 F. I have practiced both winter and summer protocols I am suggesting in this book and our Nigerian Dwarfs have done fine.

Remember that temperature is not the only climate condition factor. As with most goats, moisture is very important to manage. As with the Pygmys, if the ground is muddy, give them some planks to walk around on to avoid the wet on their hooves. They will use and appreciate them. Make sure they have shelter from the rain.

Prone to any illnesses or conditions?

Nigerian Dwarfs are hardy and strong and not particularly prone to any conditions or illnesses as long as they have adequate nutrition, shelter, and health care such as hoof trimming, vaccines and vet check- ups.

Special Considerations

Nigerian Dwarfs breed all year. Even if you don't have a buck, your does will bleat loudly when they go into heat every month. "Bleat" is not quite the right word, "scream" is a better description. One blogger said that police showed up at the door of his house one day because someone reported girls screaming. He now goes out to sit with his goats to try to keep them quiet. They can be loud.

Nigerian Dwarf does go into heat and can breed as early as seven months, but don't let them breed until they are a year. Kidding too early can result in miscarriage, stillbirth and harm or death to the doe. Bucks have been known to be ready to breed as early as three months. If you have bucks, keep them apart from the does and make sure they don't have a chance to breed them too early.[20]

Feed

Standard nutritional needs that apply to all goats.

Health

A big plus of the Nigerian Dwarf breed is that they tend to be parasite resistant.

Talk to your vet about annual vaccines and deworming if needed.
Trim the hooves every 4-6 weeks.

Multi-functions

A Nigerian Dwarf doe can produce 2 quarts of milk a day. Like the Pygmys, the milk they produce is both higher in protein than other dairy goats as well as having one of the highest butterfat contents (6-10%) of any goat breed. This makes their milk especially sweet, rich, and nutritious.[21]

Beginner Friendly

Nigerian Dwarfs are highly recommended as a beginner goat.

Nigerian Dwarf Resources

Nigerian Dwarf Dairy Association
https://www.andda.org/

Nigerian Dairy Goat Association
http://www.ndga.org/

Search your State or Region for Nigerian Dwarf Goat Clubs

Pygora Goats

The Pygora Breed Association
https://www.pba-pygora.org/index.html

Pygora goats are a breed that was intentionally developed in the 1980s by Katherine Jorgensen in Oregon. She bred Angora goats which produce mohair fiber with Pygmy goats to produce a smaller fiber goat. The name "Pygora" is trademarked. Regardless of whether an Angora is bred with a Pygmy, the name can *only* be used for goats registered with the Pygora Breed Association.[22] If you purchase a Pygora goat, it must be registered and have papers.

General Description

The breeding did indeed produce a smaller fiber goat, but it also created a beautiful, friendly pet goat who is good with kids, can produce both fiber and milk, and does not have the health issues that the purebred Angoras do.

Height: Does 18" Bucks 23"
Weight: lbs. Does 65-75 lbs Bucks 75-95 lbs
Lifespan: 12-15 years when well cared for.
Cost: About $300.00, but the price will depend on the fiber production of the individual.

Temperament

They are affectionate, sweet, and a little less lively and energetic than the Pygmys or Nigerian Dwarfs.

Climates

All climates, but as with all goat breeds, take care in extreme cold and hot and wet/muddy conditions.

If you want the cashmere for fiber, a colder climate will make a thicker production.

Prone to any illnesses or conditions?

Pygoras are hardy and robust and not particularly prone to any conditions or illnesses if they have adequate nutrition, shelter, and health care such as hoof trimming, vaccines, and vet checkups.

Talk to your vet about annual vaccines and deworming if needed.
Trim the hooves every 4-6 weeks.

Special Considerations

While they make good-natured pets, you will also have to shear them once or twice a year even if you are not harvesting the mohair to use or sell. Pygoras are intelligent and adapt to the shearing routine easily, similar to does who adapt to being milked. Be organized with your tools, hold them, start gently and give reassurance. After a couple of repetitions, they get it and will be docile, cooperative, and without anxiety or stress.

The fiber goat section will discuss how to shear a goat and describe the equipment needed.

Feed

Standard nutritional needs that apply to all goats. If you want to increase

Multi-functions

Pygoras make wonderful pets and members of the family.

Fiber – Pygoras are interesting because they produce both mohair (the Angora's curly fiber that has a sheen) as well as the cashmere that is the underfur. The mohair is sheared while the cashmere is combed out. You need to shear them twice a year and comb them out at least once a year.

There is a third kind of fiber that Pygoras have, which is called "cashgora". This is a mix of mohair and cashmere. It does not have the sheen that the Angora fiber has. For commercial purposes cashgora is considered a lower quality fiber than either mohair or cashmere.

If you want your pet Pygoras to also serve as fiber goats, wethers are better for you as they do not put their bodily energy into producing milk. See the section on Fiber Goats for more info about caring for a fiber goat and different kinds of fibers.

Milk – 1 liter/day - high in butterfat, delicious, and nutritious.

Beginner Friendly

Yes, if you are willing to take on learning how to shear.

Pygora Goat Resources

Pygoras Breeder Association
https://www.pba-pygora.org/index.html

Nigora Goats

Pet Worlds
https://www.petworlds.net/nigora-goat/

The Nigora breed was developed after the Pygoras. By breeding Nigerian Dwarf bucks with Angora does, the first Nigora was produced in the late 1980s, and the breed was then further developed in the 1990s.

General Description

Like the Pygoras, the Nigora goats offer the cross-breeding results of "hybrid vigor". They are also a bundle of goat love. Nigora is a smaller breed of fiber goat that has the personality and friendliness of the Nigerian Dwarf. They produce even more milk than Pygoras.

Height: Does and bucks, 19-29."
Weight: lbs. They are considered a "mid-size" breed; the breed standards do not list specific weights
Lifespan: 12-15 years when well cared for.
Cost: Prices range from $150-$500. The price will depend on the fiber and milk production of the individual as well as their pedigree and whether they are registered as a breed standard.

Temperament

Loving, playful, energetic, but not hyper. A great pet.

Climates

All climates but care in extreme cold and hot and wet/muddy conditions apply as it does to all goat breeds.

If you want to harvest fiber, a colder climate will make a thicker production.

Prone to any illnesses or conditions?

Like the Pygoras, Nigoras are hardy and robust and not particularly prone to any conditions or illnesses if they have adequate nutrition, shelter, and health care such as hoof trimming, vaccines, and vet checkups.

Talk to your vet about annual vaccines and deworming if needed.
Trim the hooves every 4-6 weeks.

Special Considerations

While they make good-natured pets, you will also have to shear them once or twice a year whether or not you are harvesting the mohair to use or sell. Nigoras are intelligent and adapt to the shearing routine easily, similar to does who adapt to being milked. If you are organized with your tools and holding them, start gently and reassure them. After a couple of times, they get it and will be docile, cooperative, and without anxiety or stress.

The fiber goat section will discuss how to shear a goat and describe the equipment needed.

Feed

Vitamin A & D (in the winter) sun-cured hay will have vitamin D
https://animalsake.com/pygmy-goat-care

Multi-functions

Fiber – Like the Pygoras, Nigoras produce both mohair (the Angora's curly fiber that has a sheen) as well as the cashmere that is the underfur. The mohair is sheared while the cashmere is combed out. You need to shear them twice a year and comb them out at least once a year.

In general, if you want fiber goats, wethers are better for you as they are not putting their bodily energy into producing milk. See the section on Fiber Goats for more info about caring for a fiber goat and different kinds of fibers.

Milk – 1 liter/day, excellent nutrition, high in protein and butterfat.

Beginner Friendly

Yes, if you are willing to take on learning how to shear. Even if you do not want to take on selling or using mohair, they will need shearing.

Nigora Goat Resources

American Nigora Goat Breeders Association
https://nigoragoats.homestead.com/American-Nigora-Goat-Breeders-Association-Home.html

Mini Myotonics (Fainting Goats)

A Mini-Myotonic - Victorian Farm Fainters   & Mini Myotonic Silky -Blessed Green Pastures
http://ow.ly/eXvB50CHa4F   http://ow.ly/8qYa50CHaj9

General Description

Myotonic goats (aka "fainting goats") don't faint. They are named because of a condition called *myotonia congenita.* This condition is genetic and is found not only in goats but other animals as well. To read more detail, see the links in this footnote[23]

When myotonic goats are surprised or scared, they have muscle spasms resulting in them collapsing and being unable to move for about 10-15 seconds. The episode is painless and doesn't seem to disturb them. I've been told that some will even continue chewing their browse during their "faint" until the muscles release so they can walk again.

Height: 17-25"
Weight: lbs. 55-60 lbs.
Lifespan: 10-12 years average, as long as 15 years.
Cost: Typical prices for registered mini myotonics are $375 bucks $425-$500 for does

Temperament

Myotonic's condition makes them more docile. They are also intelligent. They are bred to be pets and many people love them.

Climates

All climates but care in extreme cold and hot and wet/muddy conditions apply as it does to all goat breeds.

Prone to any illnesses or conditions?

- Apart from the weird myotonic condition, they are hardy. In fact, they carry some parasite resistance[24]
- Talk to your vet about annual vaccines and deworming if needed.
- Trim their hooves every 4-6 weeks.

Special Considerations

Mini myotonics are wonderful for children, but make sure that the children are taught not to frighten the goat repeatedly for "fun" just to see it "faint". This is inhumane and can cause a goat to become stressed, anxious and nervous. While the actual episode is not harmful or painful to them, if you identify yourself as a source of sudden fright, the goat will not trust you. As with any animal, distrust, anxiety and stress often leads to behavioral or physical problems. If you want a loving, gentle pet, be kind.

Feed

Standard nutritional provisions apply.

Multi-functions

Most mini-myotonics are bred for being pets. You can get other breeds better suited to other functions, including Tennessee Fainting Goats who are raised for meat.[25]

Beginner Friendly

Yes.

Myotonic Goat Resources

Myotonic Goat Registry
https://myotonicgoatregistry.net/

Here is an excellent history of the myotonic goat breed
https://www.hobbyfarms.com/myotonic-goats-3/

These breeders sell registered mini-myotonics for pets.
http://victorianfarmfainters.com/fainting-goats/
https://www.hobbyfarms.com/myotonic-goats-3/
http://blessedgreenpastures.com/
https://blackwalnutfarmtn.com/faintinggoats/

Full-Size and Mini LaManchas

Wide Open Pets
https://www.wideopenpets.com

Even though LaManchas were bred in the U.S. specifically as superior dairy goats, it is their temperaments that makes them a fine recommendation for pets. The mini LaMancha came about by breeding a Nigerian Dwarf buck with a LaMancha doe.

See the dairy goat section for more detail about their history.

The mini LaManchas can also be registered with the Mini Dairy Goat Association or the Miniature Goat Registry. See this footnote for the details of percentages and a description of how the breed is kept to a breed standard.[26]

General Description

The first thing you notice about a LaMancha is that they have very small ears. The "gopher ear" is so small it's hardly there at all. The "elf ear" is no longer than 2". From a distance, it looks like they have no ears at all.

Height:
- minis- does 27" bucks 29"
- full size: does 28" bucks 30"

Weight
- Minis- 90-120 lbs

- Full Size: does approx. 130 lbs, bucks 165+ lbs

Lifespan 7-10 years

Cost:

- Minis: $75-$100 unregistered, $200+ registered.
- Full size: $250-$400 depending on buck, doe, wether, age and registered or unregistered.

Temperament

LaManchas are affectionate and love being touched and petted. They will follow you around and rub up against you for pets and scratches like a dog. They are intelligent, and sometimes described as "pet like" and having a desire to please their humans.[27]

When I visited another goat farm recently, I entered the paddock and it was the LaMancha that came up to me immediately, asking for pets and wanting to be near me. They are truly extraordinary in this way.

LaManchas are calmer and move slower than the breeds we've mentioned earlier. As adults they don't tend to chew on you when you walk in the paddock, they just greet you lovingly.

Like many goats, LaManchas can use their intelligence to be excellent escape artists, particularly if bored. Make sure your fencing is adequate to keep them in and predators out.

Climates

LaManchas are hardy in cold climates but have very short coats so they will need protection. Also, the ears have a pro and a con for cold climates: pro – they are not subject to frostbite as the ears of goats such as the Nubian. However, the fact that their ears do not cover the ear opening means that they need to be protected from wind, rain and snow.

Prone to any illnesses or conditions

Aside from any complications from the ears getting exposed to wind and precipitation, they are hardy.

Feed

As a pet they have no special feed considerations.

Multi functions

Their main function was originally as dairy goats, but they are listed here as a recommended pet because they are so affectionate and love human company.

Dairy goats – produce 2/3 – 3/4 gal of milk a day. They can give you 2-3 quarts of milk a day without being bred for 2 years.

Beginner Friendly

Yes

Resources

LaMancha Breeders Association
https://www.lamanchas.org/

The American Dairy Goat Association
https://adga.org/

Miniature Dairy Goat Association
https://miniaturedairygoats.net/

The Miniature Goat Registry
https://www.tmgronline.com/registration

National Miniature Goat Association
https://www.nmga.net/

Excellent explanation of breeding mini LaManchas here
https://sweetgumminifarm.webstarts.com/minis.html

Brush Goats

All goats are brush goats! Brush is what they naturally eat. Goats are "browsers" vs. sheep, cows, and horses who are "grazers" of lower growing plants and grasses. If you specifically want to clear brush on your land or if you're going to maintain a herd that you can rent out to other people, then a herd of goats for clearing brush is a great asset.

Even though eating brush is the most natural form of food for any goat, when humans use goats as a working animal to clear brush, there are considerations. We will discuss:
- Which goats make the best brush goats, and which are **not** recommended for brush clearing
- Plants that are toxic to goats
- Fencing for paddocks
- Protocols for your brush goats

Best Goats for Clearing Brush

As with any domestic species (such as dogs, cows, horses, or sheep), there are registered animals that meet the breed standard for purebreds, and they are more expensive than those who do not meet the registered breed standards. Healthy, wonderful goats are sold as "culls" by breeders who do not want to breed a goat whose characteristics might be merely cosmetic. For example, a breed may sell LaMancha kids as "culls" because their ears are too long, and they cannot be shown or bred as a breed standard. You can get some exceptional goats for low prices by checking with reputable breeders for their culled goats. The warning here is that you want to make sure the goat is not being culled because it is diseased.

You can also get hybrids for brush clearing. The term "hybrid vigor" was coined for a reason. Hybrids are hardy. They are often more resistant to parasites, have good temperaments, strong builds, and are climate hardy.

You can definitely get healthy hybrids or purebred culled goats but do your research and make sure that the seller is not unloading a goat that may bring illness into your herd or turns out to have a condition that results in having to put them down. Several mentors have told me independently that sellers on Craig's List and at auctions are often selling goats who are ill or disabled. The best way to acquire culled goats who are healthy, and a "great deal" is to familiarize yourself with the breeders and arrange to take any cull from goats they may have already or may be born in their next round of kids. Make sure they earn your trust, get reviews, and local opinions if available. Your vet may also be able to help with recommendations.

Another critical factor is that if you get a goat that has been feeding in a paddock, make sure that they have been browsing vegetation and not just raised on commercial feed (aka dry lot).[28] You want your goats to be happy being set out for brush clearing, if they have been raised on dry lot they are unlikely to be suitable as brush goats.

As you consider which goats you want for your herd and which goats you may have currently[29], consider their sizes. Varied heights are useful if you can arrange it, the smaller goats will take care of the lower brush, and the taller ones (especially Alpines) will even stand up on their hind legs to get to higher leaves.[30]

Here are a couple of specific breeds to consider either as purebred culls or hybrids.

Alpine Goats (See Dairy Goat section for more detail)

Alpines are large and will tend to take on leadership in a herd. They also tend to be excellent brush clearing goats because they eat up the brush at a fast pace. As mentioned, they will also literally climb trees to get to what they want as well as stand on their hind legs to do your tree trimming.[31]

Alpine goats were bred as dairy goats. If you are using an Alpine as a dairy goat, don't use her to clear brush, her udder may get damaged. A wether Alpine goat is an excellent choice for brush clearing.

Pygmy Goats (See Pet Goat section for more detail)

Those little Pygmy goats are great munchers. They will get the lower stuff while the Alpines and Boers get the higher vegetation.

Angora Goats as a Special Case for Brush Clearing

If you plan to raise Angora goats for mohair anyway, you might turn them out for brush clearing after shearing in the spring. Angoras will eat up docks, thistles, nettles, and other weeds. This is the only advantage they have for brush clearing; you will see them listed in the goats to avoid in the section below. However, if you do have docks, thistles, and nettles that need work, and you are going to have Angoras for fiber anyway, this could be a strategy for you.

Boer Goats (See Meat Goat section for more detail) Not recommended for Beginners

Boers are large meat goats. They are fast brush clearers, and you can also find good crosses with dairy goats. If you want to raise goats for meat, this would be a good choice, but not as a beginner. Boer goats are often described as "gentle giants" and some will tout their "docile" nature. This may be true, but when an individual is challenging to train, they are large and very powerful. They can and will protect themselves. This is a good trait for a goat who is out in the wild brush clearing, however, the downside is that if they decide they are angry with you, then you have a large, heavy, muscley, formidable animal to contend with. If you purchase an adult, make sure that they are docile, will allow you put on a harness, lead them, and allow you to look at their hooves and examine them. If you purchase a kid, get the best advice on training from the breeder or a local expert.

Again, if you want to get goats for brush-clearing, don't start with Boers, but add them in later as "power brush clearers" if you choose to.

Goats to Avoid for Brush Clearing

- Active milking goats. This might seem obvious because it would be hard to milk a goat twice a day if she was out for brush clearing. There is another reason, though; the udders are prone to scratching and injury when out in the brush.
- Any dairy goat who may have had an injury to her udder in the past, even if she is not milking now.
- Goats with horns – the horns can get caught in brush as well as in the electric fence. Use only polled goats for brush clearing unless you supervise them at all times.
- Bucks – especially when they are in rut.
- Any adult goat that is not experienced or comfortable riding in a trailer *or* who shows a lot of resistance being led into a trailer.
- Any goat that is not easy to catch and lead
- Any goat that is not docile when examining hooves, mouth, and body.
- Bring your trailer with you when you go to a seller to interview a prospective brush goat. Besides health, you are looking for a cooperative animal who will be easy to catch, lead, and ride in a trailer. Kids will be learning, and that's fine. Adult goats may not have had experience in a trailer, but with assurance and guidance, they will get used to it. If the goat you are interviewing shows a lot of resistance getting in, it is probably not a good choice for your brush goat herd. This is especially true of the larger breeds such as the Boer and Alpine, but even a stubborn Pygmy or Nigerian Dwarf can become very difficult to move if they don't want to.

- Angora goats are not recommended as a choice of goat to acquire for brush clearing because they are sensitive to wind and rain (even though they have thick mohair, it does not have the waterproofing lanolin). They are not hardy when they are wet. Also, their beautiful long curls get tangled in the brush. Your fiber will be damaged, and you will have a big job brushing out the debris that will stick in the mohair.
- Even so, as we mentioned above, if you are raising Angora goats *anyway* for fiber, then they can be an asset out in the brush after shearing every year. They really like the weeds, thistles, nettles, and docks, so using Angoras can complement the brush clearing accomplished by the goats who prefer the woody bushes.

What Brush Goats Cannot Eat

Goats can eat a lot of things, but not all. Some plants are toxic. Also, there may be plants in your particular ecosystem that may make them sick. Talk to your vet about any warnings for your specific area. Below we discuss plants that are toxic to them as well as brush goat protocols that will help you keep your goats healthy as they munch away at the brush you want to clear.

Sometimes it can be confusing to check about the safety plants online because you can get conflicting information. This is another reason to check with your vet or local goat experts.

Short List of Plants That Are Toxic to Goats

Trees that are Toxic:
- Yew
- Stone fruits (e.g., Peaches, plums, apricots, or cherries)
    - The leaves of stone fruit trees can be poisonous to goats, especially when they wilt. They gradually accumulate a compound containing cyanide. As it intensifies, it can poison your goats. [xxi]
- Oak
    - Some brush goat owners do not allow their goats to eat oak leaves at any time, and others say it is ok except for the spring. It is worth noting that it is the spring when they are most toxic to goats. Oak leaves can cause kidney damage.

Bushy plants that are Toxic:
- Bracken (fern)
- Rhododendron
- Oleander
- Azaleas

Other plants that are toxic to goats:
- Iris
- Buttercup
- Rhubarb leaves
- Mountain laurel
- Lambs-quarter
- Pokeweed
- Poison Hemlock

A more extensive list can be found on Cornell University's Department of Animal Science website, the North Carolina Extension, and explanations about the toxic substances and how they work in the body of the animals.[32] That same footnote includes two more lists of plants that are deadly or toxic to goats. Again, advice from your local vet who is knowledgeable about brush goats and dangerous plants in your area is recommended as well as advice from other brush goat owners in your area.

Fencing for Brush Goats

A goat owner has to think about fencing for **any** goat, but for brush goats, there are special considerations:

Goats are super intelligent. Smart enough to *not only test* an electric fence but to *come back and test it again* to make sure it is still working. They can succumb to the "grass is greener on the other side of the fence" syndrome and become a bit obsessed and driven to get to an area outside the fence where they think there is better browse.

Brush goats are often more vulnerable to predators because they are away from the house where there might be livestock dogs and human guardians.

On your own land, install posts to support an electric fence that contains an area, which will then be moved as you rotate them through the areas you want to be cleared. Suppose you have more than 5 acres of land. In that case, you can put a permanent fence around the perimeter of the larger area and set up posts to accommodate moveable electric fencing in smaller areas to rotate them. This method is convenient and also gives the protection of double fencing.

If you are renting out your brush goats, the brush goat provider typically provides the fencing, which means you will need the equipment to put in post holes for the supports as well as have moveable electric fencing.

We recommend electric *wire* fencing rather than electro*net* fencing as goats can get their heads stuck in the net, pull it down, subject themselves to repeated shocking or even electrocution.

There are many resources online to show you how to set up electric fencing properly; I especially like this tutorial from a brush goat expert. [33]

Sometimes people ask about tethering their goats by tying them to a stake and allowing them to move around an area. This is not recommended for two reasons:
- They can get wrapped around the stake or tangled. They have been known to even strangle themselves after getting trapped and then panicking.
- They are incredibly vulnerable to predators; there is nothing from keeping the predator out and away from the goat, and the goat can't even run and have a chance.

I will say that I use a *form* of the tethering method sometimes. There is a meadow in front of my house that needs clearing in the spring. The last couple of seasons, I've brought out my goats and tethered one by a leash to a "zip-line", aka "high-line" (a cord running between 2 trees that is about 6-7 feet high). I like a zip-line instead of a stake because the goat will not get tangled, nor will she entangle others. Our herd is well settled; the zip-lined goat is the lead goat in the herd. The other goats will stay close to where she is. *I never leave them there alone.* I pull out a camp chair to read, bring my laptop and wifi hotspot or just sit in the "zen" experience as the goats browse away at my wildfire risk. We have mountain lions, bears as well as local dogs; I am well aware that my goats need protection.

We installed posts around the rest of our land so that it is easy and routine to set up paddocks with temporary electric fencing. You can get 100-foot industrial-grade extension cords. If your goats are clearing areas that are further away from an electrical source, you will need to provide solar-powered electric fencing. Here are some examples as well as reviews for solar chargers for electric fences. [34]

Protocols for your Brush Goats

It is always recommended that you set your brush goats out to clear in the spring and early summer, but especially if you want to eradicate plants from an area completely. Setting your brush goats out in the late summer or autumn will bring down that year's growth but will have minimal effect on the growth the next year. If your goats are set out to clear in an area every year in the spring or early summer, you can eradicate plants entirely over time.

Another reason for letting them out in the spring and early summer is that the plants have the most nutrition for the goats at this time of year. They will seek their nutrition from trees or escape if they do not get what their bodies need from the brush.

Think about how and where you are going to rotate your goats. They will not be happy if they feel bored with the food or feel that it's running out. They will become destructive and start eating the bark off of trees. Goats can eat off the bark around the trunk of a tree, and it can die in weeks. Become a keen observer of your goats to make sure they are happy and content in their clearing paddock. Notice whether they are showing signs that they want to move on. If you find that you want to do more clearing to an area after the goats have left, you can always bring them back once the plants grow more green leaves.

If you have other livestock, sheep, cows, or horses can be mixed or rotated with goats for land clearing. The "grazers" clear the grasses while the goats clear the brush. If you have chickens, they will keep the pests down resulting from the livestock's manure.

Not only timing but the *placement* of your rotation is apparently important to goats. They prefer to work from the bottom up on a slope rather than from the top down. I expect this might be because the lower part has more water. Whatever the reason, my observation living on a mountain is that they will be more content if you make your first paddock be the lower end of the clearing area rather than the upper end. This can be to your advantage for fencing as you'll have a clearer area for moving the fence up the slope after the goats have cleared the lower space.

Figuring out how many goats you need per acre and how long it will take for them to clear it is tricky as it depends on their size, breeds, and how aggressive they are in their browsing. Obviously, it also depends on the brush. Is it dense or sparse? A couple of "thumbnail" example results are:
- 10 goats for an acre takes about a month to clear
- 30 goats on an acre (this is about the maximum capacity) takes 1.5-2 weeks

You will have to experiment and get a sense of the kind of brush you have and how quickly the specific goats you have eat it up.

Caring for Your Goats While They Are Brush Clearing

Nutrition

Goats will be eating weeds, native grasses, leaves, woody stems of bushes and trees.

Make sure they have plenty of fresh water. Many goats will get dehydrated rather than drink dirty water. In areas where we cannot drive up to the paddock on our land, I put 5-gallon water containers in a cart and wheel them over two by two. Observe their water drinking to make sure they have enough.

Supplement their browse with some grain and their usual minerals as well as access to sodium bicarbonate (aka baking soda). Baking soda helps maintain their pH balance, and it is also helpful to reduce mild cases of plant toxicity.[35]

Watch your goats for any signs of acute toxic reaction. Prevention is always the best. The tricky thing about toxicity is that they may not be symptomatic for months or sometimes years. This is true of not only toxic plants but herbicides as well. Be vigilant about the area where you are letting your goats browse.

Shelter

They need to be able to get out of the rain and wind and have ample shade. Some brush clearing goat owners rig their trailers to be a portable shelter during brush clearing.

Fencing + Extra Protection

It is critical to keep predators out. If you will keep your goats out for extended periods of time or offer goat rental for brush-clearing, consider getting a livestock guardian dog. Livestock guardian dogs (aka LGDs) are bred to be closely bonded to their stock and extremely protective. They are huge. Do your own research about them. They were bred to be a solitary protector for herds in harsh conditions for extended periods of time. We have had three of them; the one that is now protecting the goats is incredibly bonded to them. The newborn kids will snuggle up to him like he is their mother! He is patient, kind, and protective. They must have the proper training. If not trained and left to their own devices, they can become over-aggressive to humans and other dogs. LGDs can be a great addition to any goat herd protection plan, but all the more so for brush goats.

Resources

Lists of Plants that are Poisonous and Deadly to Goats:
- *https://www.tractorsupply.com/tsc/cms/life-out-here/the-barn/animal-medication-for-goats/goat-care-and-poisonous-plants-to-goats*
- *https://backyardgoats.iamcountryside.com/feed-housing/poisonous-plants-for-goats-avoiding-dastardly-disasters/*

Livestock Guardian Dog Resources
- https://en.wikipedia.org/wiki/Livestock_guardian_dog
- https://www.thespruce.com/choosing-a-livestock-guardian-dog-breed-3016777
- https://morningchores.com/farm-dogs/

Reputable Livestock Guardian Dog Breeder
- https://coloradomountaindogs.com/

Dairy Goats

Humans have been raising goats for their milk (and meat) for thousands of years. Goat milk on its own is sweet, rich, and nutritious, and homemade feta cheese will make your eating skyrocket into the gourmet realm.

If you want dairy goats, how much milk will you need for your milk and cheese? The last thing you want to do is purchase multiple goats that suddenly have to milk twice a day if you have no way to keep up with their production.

Dairy goats are a big commitment. Does your lifestyle accommodate goat milking twice a day? It might be totally doable to milk a goat before and after work, or it might be a burden. Remember that you have to commit to milking them in any weather. Also, they must be milked if you are busy and even if you are sick. Many people raise a couple of dairy goats as a family project, so the milking does not rely on just one person.

No matter what breed you choose, consider that a higher cost for a healthier animal may well be worth it for the milk production you will get. If you want dairy goats, pinching pennies on your goat purchase will not save money in the end.

If you know a local friend or mentor who would be willing to come with you to visit a reputable breeder, that would be ideal. They can show you "hands-on" how to handle a dairy goat and check for their temperament as well as health. Throw in an offer to take them out to lunch and have a conversation after you visit the breeder. This can be the best resource you will find.

Nigerian Dwarfs (see the pet section for more breed detail)

Milk quantity:1-2 quarts/day
Butterfat: 6%
Protein: 4.4%

If you want a dairy goat or two for the purposes of milk or cheese for your family, then the best breed for this purpose is the Nigerian Dwarf.

Nigerian Dwarfs produce about 1-2 quarts of milk a day that is high in butterfat and great for cheese. As we said in the pet goats section, they are friendly and easily handled.

As dairy goats, one of the best things about them is that they breed all year instead of just during one season. The does need to be bred to keep their milk production up, so you can choose to stagger their breeding in a way that ensures you always have a goat to milk but don't have multiple births and tiny babies all at once.

Alpine Goats[36]

Milk Quantity: 3.5-4.5 quarts a day
Butterfat: 3.4%
Protein: 2.9%

Height Does 30" Bucks 31-32"
Weight Does 135 lbs Bucks 170
Lifespan: 8-12 years
Cost: $100-$300

If you need more milk than the 1-2 quarts the Nigerian will give, then an Alpine goat might be a good choice. Alpines are medium to large goats who are generally very hardy.

If you are a beginner who wants this breed as a dairy goat, make sure you interview the breeder carefully about the goat you are getting. Ensure that you can handle it, pick up its feet, look at the hooves, give it a check over, and see if it will follow you. Also, find out if it is trained to work with a halter and be led.

Temperament

Alpine goats are smart, curious, and friendly. Sometimes they can be stubborn, and they are a medium-large size goat, so a significant animal to deal with. Apparently, individuals vary because I see people on goat forums swearing that their Alpine is sweet, docile, and never shows stubbornness. I've not met one like that. The Alpines I've met are described as sweet, smart, and fun, but when they don't want to do something, they don't want to do it. Alpine owners I know experiment with tricks on their goats to get them to break out of the "stubborn moment" without pulling, tugging, or immense frustration.[37]

Climates
Adaptable and hardy in all climates

Prone to any illnesses or conditions
None

Special Considerations

Choose well: ask questions and check for a cooperative nature. Train well: If you are getting a kid, then you have the chance to raise her to be trusting and cooperative from the start.

LaMancha Goats (see Pet Goat section for general breed details)

Earlier, we met the LaManchas in the pet goat section. Their affectionate personalities and docile natures make them a joy as a milk goat. Here is more information about them as a dairy goat:

Milk Quantity: 2-3 quarts of milk per day for 2 years before having to "freshen" (breed again to encourage production)
Butterfat: around 4%
Protein: 3.2%

The fact that they only need to be bred every two years means that you could have two does and alternate breeding one each year.

The breed originated in the US in the 1920s by Phoebe Wilhelm. It was not until 1958 that they were accepted as a registered goat breed, and the stock has been bred and developed since then to maintain a breed standard.

LaManchas may be culled by breeders just because their ears are too long for the breed standard. If you do acquire one of these goats, your kids will not qualify for registration and their kids will have a much lower cash value than registered goats. Generally, over time it is cost saving to pay for a registered goat and have the option of selling kids that command a higher price when they are born.

Saanen Goats[38]

Saanen goats were originally bred in Switzerland. They are gentle, friendly, and cooperative.

Butterfat is highly prized among many dairy goat raisers because they want rich milk that is delicious and will also make cheese. However, some people prefer a lower fat content in their milk and are not interested in making cheese.

If you would prefer milk that is more like cow's milk in richness, the Saanen may be a fine dairy goat choice for you.

Milk Quantity: 4 quarts a day

Butterfat: 3.3%
Protein: 2.9%

Height: Both does and bucks are 31-½".
Weight: Does 123 lbs; Bucks 187 lbs
Lifespan: 15+ years
Cost: A doe is around $300.00 from a reputable breeder

Temperament

Saanens have one of the finest temperaments of any goat. They are docile, gentle, and wonderful with children. If you have the capacity to use 4 quarts of milk a day, a Saanen is a strongly recommended breed as a starter goat.

Climates

They are hardy in any climate, but they need lots of shade on hot, high UV sunny days as they can get sunburned and even get skin cancer easily.

Prone to any illnesses or conditions

Can be prone to skin cancer if not protected.

Special Considerations

Their Swiss origins made them agile and excellent jumpers. The Saanen is a goat breed who will literally climb trees. They need fun places to hop on and get up high, but make sure it is well away from your fence.

Other Dairy Goat Breeds

There are many dairy goat breeds; the four listed above are common and recommended for beginners.

Here is a brief description of two more breeds that may interest you:

Toggenburg Goats
These are the oldest known dairy goat breed from Switzerland. Medium-sized, not known for great production. They are friendly but very stubborn.

Nubian Goats

The Nubian goats that are the most common are actually "Anglo-Nubian" goats. They were crossbred in Great Britain in the 1920s from Middle Eastern Goats and Old English Milch goats. [39]

Nubians are solid producing dairy goats with a high butterfat content, so their milk makes great cheese. They are also friendly, affectionate, and *can* be a real joy. They are not on the top of the beginner's list because they are vocal, very loud, and can be impossibly stubborn.

Equipment for Dairy Goats

A compliant goat who can be haltered and led to the milking stanchion

Milk stanchion with grain bucket

A milk stanchion is a "milking station". It is elevated so you can milk the goat without being on the ground and holds the head of the goat so she can't back out. Here is an example of a beautifully made wooden stanchion. The link will show you this and 14 more DIY plans that are free.

Plain Spin
https://www.planspin.com/goat-milking-stand-plans.html

In the next footnote you will find a number of sites that sell homemade stanchions as well as metal ones from larger companies. Bear in mind that a metal stanchion is recommended if you are going to be using raw milk, as wood is porous and absorbs bacteria. Wood is much more difficult to keep clean. [40]

A "hobble": this is to keep the goat from kicking you. I love this particular one, made by a goat raiser who knows how to make a hobble that is both comfortable and effective. A hobble is a useful thing to have around for health exams and necessary procedures as well.

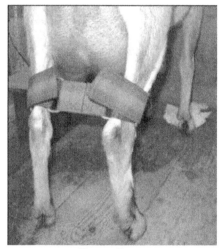

Fiasco Farm
https://fiascofarm.com/goats/hobble.htm

A Large Stainless Steel cooking pot with lid

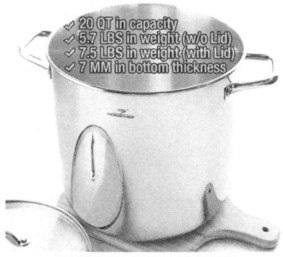

COMMERCIAL GRADE HEAVY DUTY
Amazon
http://ow.ly/ovz850CHazC

Goat brush to get hair off to keep it out of the milk

Amazon
http://ow.ly/jgVm50CHaK8

Bucket with mild soap and 2 wash clothes for washing teats and udder
- It is critical to avoid germs transmitting from one teat to the other.

- Use one cloth to wash one teat, then put it aside (not back in the water) and wash the other teat.

A yogurt container or equivalent. Give a couple of squirts into a container (used only for this use) to make sure that the teat is open and didn't have any dust/dirt, etc. hiding inside. The first couple of squirts have the most bacteria, so it's good to keep it out of your milk and equipment.

A stainless-steel strainer with a fine filter to keep out hair and small flecks of dirt, straw, hay and poop.

Large stainless-steel stockpot for pasteurization
Food thermometer
Lots of ice (details of pasteurization process below)
Or
A Pasteurizer (See example of a pasteurizer below)

How to Milk a Goat

There are many videos and descriptions of goat milking procedures. We recommend that you watch these videos.[41] Using the list of equipment above, you can follow these steps:

1) Sterilize your stainless-steel milk bucket or stockpot
Stainless Steel does not absorb bacteria and pathogens, and there are no seams or crevices where microorganisms can hide.
- Sterilize by filling it to 2/3 full of water and bringing it to a rolling boil. Cover it and let it maintain a rolling boil for 5 minutes.
- Be cautious when you take the lid off and pour out the boiling water. Put the lid back on.
- *Keep the lid on* when you go out to the milking stand

2) While your water is boiling in your pot, prepare warm soapy water with two washcloths. You can use a plastic bucket for this; make sure that you use a gentle soap like Dr. Bronner's baby soap. You don't want detergent, bleach, fragrance, or harsh chemicals in your milk or on your goat's skin.

3) Get your sterilized pot and warm wash water and washcloths out to the stanchion, then....

4) Find the goat! Lead the doe to the milking stanchion. If you give her a little grain to keep her still, then have that ready in the stanchion. Make sure your hobble is "to hand" if you need it.

5) Use the goat brush to get as much hair away from the milking as you can.

6) Wash the teats and udder. You have two washcloths because you don't want to wipe the bacteria from one teat onto the other. Wipe off one teat and ½ of the udder with one washcloth. Do NOT put the washcloth back in the water. After the first, then use the other washcloth to wipe off the other teat and the other ½ of the udder.

7) Place the sterilized pot under the goat, squeeze the teats, and get your milk. The videos in this footnote demonstrate the milking technique – you're not "pulling" as much as squeezing in a circular fashion.[42]

8) When milk slows down, remove the bucket, put the lid on (so hair doesn't fall in), and tap the udder to stimulate the flow. It doesn't hurt the doe; the kids do this much more aggressively than you will need to.
Put the bucket back and keep milking until the milk slows down.

9) Keep milking till the milk slows down again

10) Put the lid on the pot and *get the milk in the refrigerator ASAP* so you avoid that "goaty" taste.

This series of steps has worked well for me. When I first started it seemed awkward and I had to think about each step, but very quickly it became a simple, efficient system.

Do I Need to Pasteurize the Milk, or Is Raw Goat Milk Safe to Drink?

This is a hotly debated question; we will leave you with a summary of those who argue pro and con. You will also find resources to read, watch, so that you can make your own informed decision.

The "pro raw milk" advocates make these points:[43]

• Pasteurization kills not only any harmful bacteria but all the good ones as well. Our guts need good bacteria, and raw goat milk can be a source for this.

• Raw goat milk has an enzyme called lactase that aids in the digestion of milk. Pasteurization kills off that enzyme, so you are without the natural digestion aid inherent in the raw milk.

- Another enzyme that is destroyed by pasteurization is phosphatase. This enzyme enables the absorption of calcium. When the pro pasteurization advocates argue that there is no nutritional difference, that is true – there is not *less* calcium; however, your body will not be able to absorb the calcium that is there because phosphatase has been eliminated.

The "pro pasteurization" advocates make these points:

Zoonotic diseases are those that can be passed from non-human species to humans. They can come from wild or domestic animals. Raw goat milk can result in zoonotic diseases such as salmonella, E. coli, listeriosis, and chlamydia. Pasteurization is proven to be an effective and safe way to eliminate these risks.[44]

There may be raw milk dairy farms that can sell their milk safely. Still, dairy goat owners must ask themselves whether they can realistically have (or want) the controlled environment for feeding, milking, and goat healthcare required to maintain this standard of safety.

It is worth noting that some states allow commercial raw artisanal cheeses that have been aged 60 days to be sold on the market. Others do not, check your state laws if you intend to sell goat cheese.

Do your research about the safety of raw goat milk and any legal restrictions in your state. Some states do not allow you to sell it; others will not allow you to give it away. Note that even many raw milk advocates say that those with immune disorders or pregnant women should not drink raw milk.

A word of warning: I have found some raw milk advocates who will make claims about the FDA or CDC reporting either that data proves it is safe or that they support raw milk consumption. When I clicked on their link, the CDC or FDA article said no such thing. Don't take any claims for or against on face value; follow their reference and read it yourself to ensure that it is credible so that you are making a decision based upon correct information.

You need to form your opinion about the safety of raw milk. You also have to decide whether your lifestyle can accommodate the extra attention to hygiene that raw milk production requires.

Raw milk production requires rigorous milking and processing protocols. There are other standards of hygiene and pathogen prevention. We will start with the general upkeep protocols:

- Clean out the paddock every day. Goats will lie down in their poop and expose their udders to bacteria and pathogens.
- Change the bedding a couple of times a week

- Make spaces for the does to lie down that are raised off the ground and away from their poop in the pasture.
- If you have livestock guardian dogs in the paddock for protection, remove their poop as soon as you possibly can – more than just once a day. You can also train guardian dogs not to poop in the paddock and give them regular walks to relieve themselves as you would a house dog.
- Keep your milking stanchion clean. A metal stanchion is more manageable to sterilize than wood if you are drinking raw milk. If a bird poops on the stanchion, wash it off and sterilize. Birds (including chickens) can carry zoonotic diseases and pass it to you through contact with your equipment or through some infections that get passed to goats.

In addition to some of the practices listed above, The Micro Farm Project offers clear, step by step instructions for producing raw milk as safely as possible. If you decide to produce raw milk for your family, we highly recommend the practices outlined in this resource.[45]

How to Pasteurize Your Milk

Pasteurizing your own milk at home is easy and straightforward, but it takes about 40 minutes per gallon if you do it by hand. Watch this excellent video that shows you step by step how to do it. The steps below are a summary of her instructions.[46]

- Put a gallon at a time in a stainless-steel stockpot.
- Put on medium heat, stirring often, so it doesn't stick and burn on the bottom
- Get it to 165 degrees for 15 seconds
- After about 10 minutes, it will get up to about 145 degrees
- Prepare sink with ice and water in the bottom – enough that will cover most of the pan when you put it in to cool.
- After 145 degrees, you will need to stir more often, and towards the end, you need to stir constantly.
- After the 15 seconds at 165, you want to cool it as quickly as possible.
- Place the pot into the ice and water in your sink. Continue to stir so that it cools off evenly
- After 5 minutes, it is cooling down – add more ice if the water is not cold
- You want it to cool down to between 65-55 degrees
- It takes about 20 minutes for the gallon of milk to get to 60 degrees when stirred constantly
- Pour into sterilized glass jars for storage and refrigerate
- Watch the video to get all the details.

If you don't want to take the time to do that process every day, you also have the option of purchasing a home pasteurizer for $690. If you have the budget for this expenditure, it will save you a lot of time and may well be worth the investment.

Amazon http://ow.ly/8xYk50CHaSX

Fiber Goats

Goats have served multi-purposes for humans for 1000s of years. Goats have been bred to provide fiber for clothes as well as blankets and insulation of portable structures such as yurts.

Most (not all) goats can produce some fiber, but there are goats who have been bred to be *especially* good as fiber goats. We will discuss a number of options you may have if you are interested in raising some fiber goats.

What kind of Fiber do Goats Provide?

Goats provide three different kinds of fiber:

Mohair is exclusively from Angora goats.
- Mohair comes from the curls of the Angora goat and has a sheen. The fibers of mohair are long, and it is sheared off.

- Mohair is used for knitted garments (usually mixed with other fibers). Hats, scarves, sweaters, gloves, and blankets are common garments made from mohair.

Note that the fiber from Angora *rabbits* is called "Angora Wool". Goats do *not* produce Angora wool. Only rabbits produce Angora Wool. Angora goats produce mohair.

Cashmere is From Any Goat That Grows a Layer of Downy Undercoat in the Winter

Cashmere goats have two kinds of fur: guard hairs and undercoat. The guard hairs are the coarse hairs that grow through the underfur and are seen on top. They are longer than the cashmere undercoat that is grown next to the skin for warmth. Guard hairs remain all year and are not shed or brushed out in the spring

- The undercoat is released from the goat and makes a pocket of air every year before it sheds.
- Cashmere has short fibers and is brushed out to harvest for use, not sheared.
- Cashmere is used in woven garments as the fibers are too short for knitting.

Cashgora is Fiber From a crossbreed Between an Angora Goat and a Cashmere Goat[47]

The Pygora goat (Pygmy and Angora crossbreed) and Nygora goat (Nigerian Dwarf and Angora crossbreed) are two registered cashgora breeds.

Cashgora fiber is not considered a high-quality fiber for commercial use. It is sometimes useful for home spinners and craft workers, but many do not want to go to the trouble of working with it. Others report that they love it. If you are a spinner, ask around to find anyone with experience spinning cashgora.

If you want to have goats as pets, for brush-clearing, or as working goats, you may want to consider harvesting the cashmere, especially if you are in a location with cold winters that will encourage the growth of a thick undercoat.

We will discuss the best breeds for fiber goats as well as the maintenance, care, and special considerations. First, let's look at what goats do *not* make good fiber goats and which ones are not recommended for beginners.

Goats to *Avoid* as Fiber Goats

The goats to avoid as fiber goats have to do with two main things:
1) Their function and
2) Their temperament

For any breed, avoid:
- Does who are breeding, pregnant, nursing, or being milked as dairy goats.
- Bucks

The basic principle is that you must decide where you want the energy of the goat to go. You can choose to send the goat's energy either to fiber or their reproductive system, but not at once. Your fiber will not be the quality or quantity you want. Get wethers for your fiber goats from whatever breed you choose. Obviously, if you are breeding fiber goats, you will have to have at least one doe and one buck. They will not be your top producers of fiber. Let their energy go to breeding. If you want dairy goats, then designate them as dairy goats, but don't expect them to produce fiber as well as milk. A body can only support so much!

There are a couple of breeds of goats that have been bred for cashmere but are very large, and densely muscled so can be a lot to handle if they feel uncooperative:
- Boer Goats
- Spanish Goats

Purebreds from these breeds are not recommended for beginners. As meat goats, they are large, strong, and powerful.

You will be learning how to comb out and shear a goat. Learn on a goat who is not easily angered. If you decide to include meat goats into your herd, then remember that Boer and Spanish goats can produce beautiful cashmere – *if they allow you to get it!*

Breeds of Goats Best for Fiber

As we said above, for any breed, the best fiber goats are wethers because their entire body energy can go into producing fiber. When you are breeding fiber goats, you obviously must choose to have at least one doe and buck, but do not expect them to be your top producers. Also, harvesting fiber when a buck is in rut or a doe is in heat, kidding, or nursing is stressful to their body. The doe's milk will not be of the highest nutritional value because she will be putting energy into growing out a new coat. A nursing or dairy goat or a buck in rut should not be your fiber goat.

Cashmere Goats

As mentioned earlier, any goat that has an undercoat can be a cashmere goat. You will see articles and listings for "cashmere" goats. It is a bit confusing because there are hybrids that are called "cashmere" goats, but it is also a description of the *function and type* of goat. Cashmere goats can be registered with the Northwest Cashmere Association and the Cashmere Goat Association, even though it is not a specific breed.

Characteristics of cashmere goats for registration are:
- "complete and consistent" coverage of fiber over the body
- "body conformation" - this is not to do with height and weight because the breed mixes vary. The specific criteria for "body conformation" are:
    - Beautiful head and horns (the horns must not be dangerous);
    - a "well-sprung" barrel rib cage;
    - and strong, well-muscled forequarters, back, and hind legs.48

Any goat that has genetics from cashgora goats is not accepted for registry as this degrades the cashmere fiber.[49]

Something to watch for is that the breeds that have been bred in Australia and New Zealand are closer to a wild goat in their nature. Their temperaments can be more challenging than the cashmere goats bred with domestic breeds such as the dairy goats.[50] These mixed breeds who are registered cashmere goats will often have bloodlines from the Spanish Meat Goats, but since they are mixed, they do not display the same aggressiveness as the purebreds. [51]

Saanen Goats

The absolute best breed for beginners who want to have cashmere fiber goats is the Saanen.

The Saanen breed is at the top of the list because they are our top recommendation for a beginner as a cashmere fiber goat. As discussed in the dairy goat section, they are sweet, docile, affectionate, and an easy goat to work with.

If you want Saanens for dairy goats, then you can get a buck, a doe, and a wether or two and get your fiber from the wethers. Alternatively, start with a few wethers, and you'll have soft, thick, high-quality cashmere from goats who will let your comb and shear them (and may nuzzle you while you do it).

If you have no goats and want to specifically start with a cashmere fiber goat, then the Saanen is the best for the mix of temperament and production.

Other goat breeds to consider for cashmere fiber goats:

Nubian Goats

Nubians can produce beautiful fiber. If you don't mind a potentially loud goat who might get stubborn on you, then they can be an excellent fiber goat, especially for an intermediate level.

If you do get a Nubian for fiber, make double sure that they are trained to be led and handled. When you get them home, make sure that you train them to allow *you* to handle them as well.

Toggenburg Goats

Being from the Swiss Alps, Toggenburgs have the capacity to produce great fiber.

As with the Nubians, these are recommended for an intermediate goat raiser. The same considerations apply.

Mini breeds for Cashmere Fiber:

Both the Pygmy and Nigerian Dwarf goats can produce a lovely, thick cashmere. If you really want fiber, you will need more of the mini breeds because of their size, but the quality of cashmere is sellable. The Nigerian Dwarfs are more highly recommended for their fiber than the Pygmys.

Mohair

The Reiland Farm
https://thereilandfarm.com/index.php/angora-goats-sheep/

Angora Goats

Mohair comes from Angora goats. It is not an undercoat; it is from the long curls on these goats. Originating in Asia Minor, they have been bred for about 2500 years.

Here is an article about a homestead that has a small business raising Angora goats for fiber. They love their Angoras.[52] It is very helpful as she shares their breakdown of cost and profit from the mohair.

We do not recommend Angora goats for beginners. If you are an advanced beginner or intermediate goat raiser and think you want to go into producing mohair, then be inspired by Iron Oak Farm in the article. Before you commit yourself to Angoras, here are some facts and considerations so that you can do your best planning to care for these beautiful creatures.

Height: Does: 36" Bucks: 48"
Weight: Does: 70-110   Bucks: 180-225
Lifespan: 10 years
Cost: $175-$550 depending on pedigree and health

Temperament:
They are gentle and docile.

Climates:
Best in semi-arid climates – hot, dry summers and cold winters. They do not do well in wet climates.

Prone to any illnesses or conditions:

Yes, they are prone to health problems; Angora goats are not hardy. They are especially prone to illness or death at birth and for 4-6 weeks after shearing twice a year. As mentioned, they do not do well in wet conditions.

Here is a list of common concerns with Angoras.
- Because of their dense coats, they are vulnerable to external parasites.
- They are more likely to get dermatophilosis, lice, scabby mouth, and flystrike than other breeds.[53]
- Their hooves are sensitive (especially to wet conditions) and easily succumb to lameness and ill-thrift.[54] Their feet issues may even result in sudden death.[55]
- Angoras are prone to infectious diseases and worms, so diarrhea is common.[56] If you don't clean the diarrhea off their coat, it will be a danger to spread in the herd as well as ruin some of the valuable mohair.

Special Considerations:
Angora goats put a lot of body energy into those ringlets. When does are breeding, pregnant, or nursing, they need to have more feed. "The hair follicles of the developing kids won't form if the dam's nutrition is lacking." [57] If your pregnant dams and does do not have excellent nutrition, you will have less fiber from their kids.

Mixed Breed Cashgora Goats

A cashgora goat is any cashmere goat that is bred with an Angora goat. The most common are the Pygora and Nigora goats we have discussed. Other crosses are with the Spanish, Boer, and Nubian goats.

The *idea* of having a goat that could provide mohair as well as cashmere without the delicate nature of the Angoras seems like a good one. The problem is that while you can get mohair from the cashgora goats, cashgora fiber is not considered a high-quality fiber.

Professional fiber goat farmers tend to shy away from cashgora goats because the fiber is not of dependable quality and labeled as "NCV" (no commercial value). [58] Even so, there are small fiber goat enterprises who are apparently very happy with their cashgora goats.

If you are considering a cashgora goat enterprise, do your research. The resources in this footnote will get you started and give you some tips.[59]

For fiber goat associations and groups, you can find websites here.[60]

Working Goats

The term "working goat" is used in varied ways. Some people use it to include dairy, meat, fiber and brush goats, and even agility and show goats. For our purposes, the term "working goat" specifically means a goat who is bred and raised to pull or carry things.

Working goats are often mixed breed goats who are bred to be stocky with long legs. They are strong and can pull carts or carry packs over varied terrain. Unless you want to specialize in pack goats and have a "string" of multiple goats in the backcountry, you can have one or two working goats in a herd of goats who are there for other purposes such as dairy, fiber, brush goats, or even pets. They are **very** handy on a small homestead when you need some help getting a cart full of something across your land and provide you with great company as you do your chores. The training component of a working goat is also an excellent activity for children as they learn to take responsibility for raising an animal and interacting well with them.

We will discuss "pulling goats" or "driving goats" who pull carts and wagons and "pack goats" who carry packs on their backs while covering distance in the backcountry.

Add: historically, goats used in mining, plowing, transportation

What kind of goats make the best working goats?

Never use pregnant or lactating goats or milking goats. They need their body energy for their kids or milk.

Bucks are....bucks. During a rut, they will not only smell but be at best impossible, and at worst aggressive. Even outside of rutting season, they are not as even-tempered as a working goat should be.

Wethers are the most highly recommended choice for pulling or packing. They will be inclined to be compliant and are bigger and stronger than does.

Pulling Goats

Goats typically pull either carts or wagons. The difference between a cart and a wagon is that a cart has two wheels and a wagon has four. The difference is not size; sometimes, carts are larger than small wagons.   Using your goat for pulling is called "driving" your goat.

Goats wear harnesses when pulling. Goat anatomy is different from horses or donkeys, so trying to use a harness for another animal is not recommended and is likely to injure your goat.

Carter Pet Supply
http://ow.ly/Ooow50CHbAR

How old does my goat need to be to pull a cart or wagon?

Your goat is not fully formed until they are 3. A pulling goat needs to be 1-2 years old when they start pulling to make sure that their bones are developed *enough* to begin to do the work. You are encouraged to start training as described below when they are kids – they can learn to be led by a collar, a bridle, and be accustomed to being fitted for and wearing a harness on walks before pulling any weight at all.

If you want to start training your one-year-old goat to pull, check with your vet to confirm that they are ready, or perhaps they need another year just pulling a small amount of weight. Individuals vary and choosing when to begin is an important decision. You don't want to end up with a suffering, disabled goat because they were pulling too early for their growing bones.

How much weight can my goat pull?

With gradual fitness training as a fully developed adult:

A doe can pull her own weight
A buck or wether can pull twice their weight

Training Your Goat(s) to Pull a Cart or Wagon

Training needs to be done incrementally, with care. You'll probably only need a few weeks, depending on how fast the goat learns. Make plans for two, 15-30-minute sessions per day.
- put some focus on your goat,
- make it fun and interesting for them,
- and be kind and reassuring.

The last thing you want is for your goat to be afraid of the cart.

It is best to train a working goat at a young age, but a goat can be trained for driving at any age, depending on their temperament and your relationship with them. Handle them, earn their trust. Get them used to being brushed, especially around the areas where the harness will be. In addition to getting them used to touch and handling, this will keep the coat in good condition and get it to lie down properly under the harness, lessening the chance of irritation.

A word of caution: NEVER expect your goat to pull anything just wearing a collar. They must have a harness that distributes the weight properly to their chest, back, and legs. If they pull weight with a collar, they can injure themselves or even crush their own windpipe.

The incremental training starts with
- a collar,
- then a bridle,
- the harness,
- then the cart or wagon.

Each step goes from
- Sniffing
- Wearing
- Walking around with it (Sometimes wearing and walking around with it are one step, sometimes the goat needs to just wear the apparatus for a few minutes a couple of times a day before being led by it.)

Here is an outline of the steps that I was taught by my working goat mentor. I encourage you to use this method along with extra information from the excellent resources in this footnote.[61]

Brush and pet the goat when you start each session. This will help you lead up to putting on the harness and the goat having positive associations with these training sessions and eventually with pulling the cart or wagon.

1) Start with leading by a collar. If the goat is not used to a collar, then let them sniff it first. They sniff, then you give a little treat, pets, and praise. Put the collar on your goat and see if you can use it to lead them around. If not, then spend 15-30 minutes twice a day with some treats to get them used to it.

See this example of leading a goat by a collar – do it holding from above, in between the ears rather than pulling from the neck.

Thrifty Homesteader
https://thriftyhomesteader.com/working-goats-your-journey-begins-here/

Start walking around using commands that will be consistent with your training and driving your goat with the cart or wagon.
- Go or Walk or "walk-on."
- Clicking sounds for "go faster."
- Left
- Right
- Whoa or Stop
- Back up
- Stand
- Some goat raisers add-in "NO" for behavior that is not welcome.

Take it a little at a time every day. The best is to do a couple of 15-30-minute sessions a day, but don't overtax them. A goat that is mentally fatigued will begin to rebel against the whole idea.

Also include praise, petting, and treats so that working on this project with you has positive associations.

2) After the goat can be led by a collar, introduce a bridle in the same way
- Let the goat sniff the bridle
- Put it on and let them walk around with it
- Lead them with it
- Begin using commands

3) After the goat is used to being led with the bridle, then introduce the harness.
- Let the goat sniff it
- I recommend tying the goat to a fence before fitting the harness.[62]
- Put the harness on and adjust it
- Lead them around with it, use the commands
  - This serves the purpose of training the goat as well as observing the harness as they move and adjusting it as needed, so it is a good fit.

4) After your goat is comfortable and compliant with being led by the harness, begin to pull a small wagon *yourself* as you walk alongside them. Goats are prey animals and get spooky when there is noise following behind them. Help your goat adjust to this by pulling a wagon along yourself or having another person come with you on your training walks. The other person can walk alongside the goat so that the goat gets used to hearing the wagon. Gradually, they can drop back further and further so that the goat is used to the sound of the wagon without being able to see it behind them.

5) When your goat is comfortable being led by a harness with a wagon beside or behind them, then hook them up to an *empty* cart or small wagon. You don't want to challenge their pulling ability; just get them used to the feel of pulling the wagon and the sensation of it being right behind them.

6) Gradually add weight in the wagon till they can happily pull the appropriate amount of weight.

Here are some resources for finding goat wagons and carts or making your own.

Here is a video that shows you how to make a goat cart
https://www.youtube.com/watch?v=eC-xvArwrnY

I love this cart made for miniature goats using bicycle tires
https://thriftyhomesteader.com/working-goats-your-journey-begins-here/

Here is a goat supply company that includes goat wagons as well as harnesses, a seat, and other working goat equipment.
https://www.caprinesupply.com/products/working-goats.html

Pack Goats[63]

Pack goats carry packs on their back rather than pulling things. They are typically used in the backcountry. Pack goats are increasingly popular with hunters, and recreational backpackers as goats are generally less expensive to keep than horses or donkeys, they browse so you don't have to pack their food, and they don't need nearly as much water as equines do. Depending on the conditions, a pack goat can go a couple of days, if necessary, without a water source.

Larger breeds (often mixed-breeds) are chosen as pack goats as they are taller and can carry more weight. Again, wethers are recommended for the same reasons as pulling goats. Doe's udders can get damaged, and they shouldn't be putting their energy into carrying things. Bucks are unpredictable at best and absolutely impossible as pack goats when in rut. Meat goats are *not* good choices as pack goats. Even though they are strong, they are built lower to the ground and are not agile. They are not very energetic either; they will not enjoy long hikes with weight and will let you know by being non-compliant.

The dairy goat breeds that are often used for pack goats are:

- **LaManchas**

Strong, great temperaments, and they enjoy being out in the backcountry with their favorite humans. This breed makes an A+ pack goat.

- **Saanens**

Their temperament and build make them excellent pack goats. Note that they don't do well packing in temperatures above 90 degrees F. Also, they can be prone to splayed hooves, so examine any Saanen that you are purchasing and make specific inquiries about their genetics. Have a look at the hooves of the parents if possible. Keep an eye on the hooves of your Saanen pack goat and watch for any signs of change in their hooves after packing. You may need to adjust the amount of weight they carry.

- **Toggenburgs**

These are built really well for packing and reportedly are very aware of anything around the trail to be aware of. They will stop and stare and let you know (quietly) that there is something to pay attention to. They are also very independent and may be stubborn when you are out with them.[64]

There are breeders who specialize in pack goats; you can make arrangements for a kid from excellent stock and lineage here. [65] My local pack goat mentor is currently breeding LaMancha and Saanen goats as a mixed breed pack goat. The hybrid vigor, temperament, and strength of these goats make an excellent pack goat. Other breeders mix other breeds. It is not recommended to get a pack goat from an auction or Craig's list or another marketplace as this is where people often offload diseased, deformed, injured, or ill-tempered goats.

- A goat breed we have not discussed yet: **Oberhasli goats**

Oberhasli goats originally came from the Swiss Alps, bred as dairy goats. They have been bred since the early 1900s in the U.S.

Height Bucks: 30-34" Does: 28-32"
Weight: Buck 150 lbs Doe 120 lbs
Lifespan: 8-12 years
Cost: $300-$500 purebred and registered

Temperament:
They are described as "calm" (which is great for a pack goat) as well as "bold". In the case of the Oberhasli, "bold" refers to the fact that they are not easily frightened and are not skittish. They are also not afraid of water. These are all wonderful traits for a pack goat. Be aware that "bold" also means that they tend to want to be "bold" with other goats and tend to want to be dominant. This is why they were not recommended as a beginner dairy goat. Oberhaslis *can* get aggressive with others in the herd and cause drama in the barnyard or, worse, backcountry. Find a breeder who has been interested in temperament as one of their breeding criteria.

Climates:
All climates

Prone to any illnesses or conditions:
They can be prone to tetanus, which can obviously be a risk in the backcountry. Proper vaccines and vet advice are strongly recommended.[66]

Can I pack with only one pack goat?

If you have no goats, you *must* get two. Goats are herd animals. They prefer numbers but need at least one other goat.

If you have a herd of goats and will only be using the pack goat for day hikes, then it's OK to just have one.

If you are going on overnights, then get two or more pack goats. The goat will be much more settled, less anxious, generally more likely to behave if they have a friend. You *absolutely* need two if you are going out for more than one night. It is in your best interest to have happy goats instead of one nervous and unsettled goat who may be less cooperative.

How old does a pack goat need to be to start packing?[67]

Since packing involves weight directly on the body, the goat must be absolutely fully grown and developed before they start bearing weight. I recently visited a pack goat breeder who said she was increasingly careful about who she was selling her kids to because a couple of people did not listen to her instructions (below) and the young goats were crippled.

They are not ready for their full capacity until age 4.

You can start practicing with a saddleback at age one, but with no weight.
Ages 2-3, you start practicing with a bit of weight, gradually increasing.
At age three years, you can start packing, but they are not ready for their full load until age four.

How to Train a Pack Goat
Below we outline some basics so that you can see the steps. These steps are based upon my own experience, my mentor's teaching, and research. Please do further research. The best place to start online is with Mark Warnke at www.packgoats.com. I have found his information to be in sync with what other experienced goat packers have taught me. His blog is referenced a number of times in the following information.

Training, of course, is best to start as a kid. We will also cover some points about training adult goats.

Training Kids to Be a Pack Goat

Your kid won't even put on the saddleback (let alone panniers) before they are age one. There is a lot of *training* to do in that first year, though.

8 weeks-1 year

The most critical thing to do early on is to bond with the kid. Give them lots of love, praise, and assurance. Establish a strong rapport and spend a lot of time with them. If you have bred a goat yourself, then you have the kids from the moment they are born. If you purchase a kid from a breeder, it is likely that you will start the process when they are about 3 months old.

Of course, do all the basic goat protocols for a kid – get them used to having their hooves trimmed, touch them on their stomachs, be comfortable with the examination of their whole bodies, and generally make them tame.

The first year you won't even have a saddle on them. They will be learning to be tied, led, go on walks, and towards the end of the first year, start giving them commands.

At about 3 mos. you can start to take your kid(s) out on the trail. Get them to follow you and to forage. As goat pack expert Marc Wamke puts it, the key lesson they need to learn is: "goats walk fast, and the leader (me) doesn't wait for lolly-gaggers."[68]

I can't stress this enough. My "online mentor" in goat packing, Marc, says that the *worst* thing you can do is to stop and call for them.[69] This has proven to be true in practice, as well. They must think you are moving forward and ignoring them. If you change the pace, do it in a sneaky way so that the goat isn't aware. If you wait for slow or distracted goats, they will learn that they are the ones in control, and you wait for them. You can walk pretty fast, even with 3-month-old kids. Pattern their brains to know that if they don't keep up, they will get left behind. This will save you a lot of trouble in the future with a 200 lb. goat who thinks that he controls the pace.

Think of their fitness conditioning as being similar to a child. First, you would take them on a trail that was easy and not too far. To have fun, they get lots of praise and stop to play some games when you get to rocks or fallen logs to encourage them to practice their agility. Let them have breaks and browse (on your terms). If you have plants on your land that are out in the backcountry, then including a branch here and there as part of their diet at home is good too.[70]

If it seems like they just need a slower pace for their fitness, then slow down your pace a little. The point is that it is *fun*, and they learn to follow you. You don't want your walks to be stressful for them or have associations similar to the child who hates P.E.

When your kid(s) has been following you for a little bit, and you stop, and they are with you, give them lots of praise and a little break – play a goat game or rest.

Gradually increase the length of the walks on an easy trail, and then advance to shorter walks on intermediate trails. Always remember that good associations and having fun, are the keys. [71]

The following timeline for pack goat training was passed down to me and most pack experts agree. For the most part it is simple common sense.

<u>1-2 years</u>

At age 1, they can start to carry a "kid pack" for practice.[72] They are still not carrying any weight for you. You are practicing your own skills at putting on a pack; your kid(s) are getting used to being "saddled up" and walking with it. They have a learning curve to get used to navigating over logs, rocks, etc., with something on their back.

Keep working with your kid and increasing their conditioning. If you are going to want to do cross-country, off-trail trekking, then take them off-trail to have to step over logs and other obstacles.

Gradually get them used to crossing very small amounts of water and increasing the challenge. Goats do not like water generally, and unless they are an Oberhasli, they tend to be afraid of it. Patience, assurance, calm assertiveness ("this is what we're doing"), and lots of praise when they cross are critical. If you have built your rapport with your kids since they were 8 weeks old, then you have built the trust necessary to start this process slowly and incrementally.

<u>2-3 years</u>

Keep using the kid packs, but they can now work up to carrying 10% of their weight.

<u>3-4 years</u>

Your goats will be ready to carry a full pack of 25-30% of their body weight.

Remember that just like any mammal, including us humans, fitness is key. It would be cruel to put a full pack on a goat for a hunting trip and then hike up a steep mountain after the goat had been kept in the barnyard for months with no conditioning. Your goat would not enjoy it, may balk and just lie down, or become dangerously exhausted. It would also make them prone to injury.

This article gives you more detail you would need to know about training your goats through the years. It also discusses winter training, which can apply to any "off-season" for packing, whatever that means for the activities you do.[73]

Agility goats

It is a bit surprising that competitive agility events for goats are only just beginning to emerge as a popular activity. Goats have evolved for agility! Generally, it is one of their greatest strengths and deepest joys. (The exception to this would be some of the meat breeds like the Boer, who have been developed to be heavy and stocky rather than agile. They can be fun as kids, but not ideal as adults for this purpose.)

Agility courses can be very simple to highly complex. It is a great practice for *any* goat owner as a game to do with your goat. Agility is a wonderful project for children as they learn to be good teachers and animal communicators. County fairs and 4H both hold goat agility events. This is a good place to observe and meet experienced goat owners.

You can, of course, purchase advanced goat agility equipment. To start, though, you can put out all or some of these things:

A stump that is large and stable enough for the goat to jump on:

Mother Earth News http://ow.ly/preM50CHbKF

Then, start adding stumps – you can make a circle or a snaky shape, or whatever shape you can imagine in your space.

Our Cool Planet
https://ourcoolplanet.wordpress.com/

And something to jump over – a couple of stumps or bricks with a long branch, broom or old ski pole. Make it only as high as their shoulder or lower.

A hoop jump made from a hula hoop, pvc tubing (I used old drip system tubing once for this purpose) or an old bike tire. This one is shown mounted on a frame – nice idea, I've always just held mine.

Pet DIYs
http://petdiys.com/diy-agility-hoop-jump/

A beginner goat tunnel
- Use a plastic bucket or other container or
- Put a tarp over a series of supports (straw bales, anything around your barnyard that could hold a tarp and make a tunnel for your goat

If you think about the concept, there may be something you can repurpose from your homestead.

Pinterest – no instructions, but you get the idea
https://www.pinterest.com/pin/334392341057548652/

If you have a child's slide (sometimes people give them away) they make great goat agility obstacles. Check that the ladder provided has solid footing for their hooves. If it does, then the goat will love climbing up.

YouTube "Spotty's Tricks http://ow.ly/lqNF50CHc8N

An A-frame of some kind is really fun and intuitive for them. I like this one using small boards so that the goat has good solid footing.

YouTube "Spotty's Tricks"
http://ow.ly/lqNF50CHc8N

Use a collar and lead, but not to *pull* the goat, to *guide* them. Patience and praise are key – make it fun so that they look forward to a little agility every day. Add it in to your time with them where you pet and scratch them and generally love them up. If you have ever worked on obedience training with your dog, then you understand that short amounts of time with a few treats and a lot of patience are critical. If you or your goat get frustrated, stop.

The first obstacle I like to use with my goats is the stump to jump up on because it is the most likely agility move they will do naturally for fun. Give a command like "jump!", "up!" or "hup!" and your goat can jump on the stump and get a treat. Another way of training them is less with specific voice commands for each activity and more just with your body and hand signals.

As with a dog, you start with the treat given *immediately* when they do the right thing, then as training goes on the treats are stretched out less and less till they are occasional.

Click on the video below to see how this young woman trained her goat. She uses some physical objects but does a lot just with her body using no equipment whatsoever. You can see her training skill as well as the fact that the goat loves both her and the activity.

YouTube "Spotty's Tricks"
http://ow.ly/lqNF50CHc8N

Note that this is a Boer kid. Clearly a wonderful agility goat at this stage, but when he starts putting on their full weight and height carried on short stocky legs, it will get more challenging to do these things! By nature Boers adults are lower energy than other goats. That goat is here to prove that at least as kids, *any* goat can be an agility goat!

Show Goats

If you are into purebred or registered hybrids, then showing may be fun for you. The glamorous fiber goats are naturals for this, but there are meat and dairy show events as well as breed-specific goat competitions. This is another popular choice for 4H kids, as the whole process teaches them a lot about tending animals in a way that is for best production and highest standards.

If you know you want to show goats, then carefully select the breeder and the goat you are purchasing. You will pay for pedigree, but if you are showing, then it will probably be worth it as you will be more successful with better genetic lines in your goat. Learn about the breed standard for the goat you are showing and ensure that the lineage of the goat you are purchasing has met these standards and proportions.

Caring for show goats is a little fussier – you will be thinking beyond health and into aesthetics for their coat and their hooves. Do research on finding the best specific diet for your goat so that their nutrition is top-notch. Ask your vet and local goat experts who also show their goats about the best products for making coats and hooves look their best.

You won't just teach them to "stand" like a pulling goat – (meaning "stand there, stay still, don't walk away with the wagon"). You will be teaching them to "stand" in order to take a position for a judge.

Here are some exemplary goats and handlers

WSU Extension
https://extension.wsu.edu/lincoln-adams/4h/livestock/

Farmtek
And an entire page of standing mini dairy goats in show

Miniature Dairy Goats
http://ow.ly/OJi450CHcVS

As with dogs, the world of show goats has its own specific culture, requirements, rules, and expectations. There are even expectations around what the handler wears. If you wear the wrong thing, you may get marked down for showmanship. Find a mentor. Internet research will prepare you to make the most of a local mentor, but there will be conflicting information. Your local 4H club is a great place to start.

This footnote offers a number of reliable resources for show goats, including a three-part series about showing goats published by Washington State University.[74]

Meat Goats

Goat meat has risen in popularity for a variety of reasons.
- Consumers are looking for lean red meat that can complement or replace heavy reliance on beef.
- Goats are smaller and less expensive than cows for small family homesteads who want to raise their own meat.
- Goats have been used for meat globally for millennia. As the population of the US becomes more diverse, the demand for goat meat has risen.

As a result, both homesteads raising meat goats for family consumption as well as meat goat businesses raising them for market have multiplied.

The best goats for meat goats are, of course, stocky, strong, and muscular. If you want to sell meat goats at a market, then you would want to focus on meat goat breeds. You can also include meat goats into your herd of dairy, brush, or other goats as a way of making use of wethers.

If you want to consider meat goats, then first find out where there is either a market to sell them or a slaughterhouse. Some areas have local meat processing that only caters to cows and not to goats. A meat processing plant designed for larger herds of cows may not accept your single goat for your family's freezer, or even several goats. Do your research to ensure that you can find a way to access a slaughtering house or a market.

We will first discuss a couple of breeds that are specifically meat goats; then, we will look at mixed breeds and multi-functional breeds that can fit into a brush or dairy goat herd.

Boer Goats

Boer goats have become the most popular and well-known meat goat in the US. They are described as calm and docile, and they have both the build for an excellent meat goat as well as the growth rate. They grow and fill out quickly.

Boer goats did not arrive in the US till the early 1990s. They were originally bred and developed in the early 1900s by Dutch farmers in South Africa.
Weight Bucks: 200-340 lbs   Does: 190-230
Cost: $200-$2000

Temperament:

They are called the "gentle giants" of the goat world. They are docile and enjoy human attention, but as adults, they are not energetic or likely to want to play games.

Climates:
They excel in hot/dry climates but will also be cold hardy with proper care and shelter.

Prone to any illnesses or conditions:
There is conflicting information about their parasite resistance. Some say they are particularly parasite resistant, and others report they have had problems. Bear in mind that the best food for goats is brush. If you raise your meat goats in a pasture like cows, they will be more prone to parasites as they will be eating plant matter next to the ground.

Special Considerations:
They will breed all year around and are very fertile.

Resources and Associations:

American Boer Goat Association
- https://abga.org/

American Meat Goat Association
- https://meatgoats.com/

American Goat Federation
- https://americangoatfederation.org/

Kiko Goats

Kiko goats are another recent breed. They were bred and developed in New Zealand in the 1980s and introduced to the US in the 1990s. They are increasingly popular in the US, but not as common as the Boer goats. There is a registry but not a breed standard.[75]

Weight: Bucks: 250-300 lbs   Does: 100-150 lbs
Cost: $300-$1500

Temperament:
Docile and easy-going.
Known to be "low maintenance" in both health and temperament.
They are good mothers

Climates:
They are both heat and cold hardy

Prone to any illnesses or conditions:
No, in fact they are known to be especially hardy because they were bred from feral goats originally.

Special Considerations:
None

Associations:

American Kiko Goat Association
- https://kikogoats.com/

Pygmy Goats

If you are homesteading for your family and want milk, dairy and child friendly playful goats around, then Pygmy goats can be a wonderful multi-purpose option. We have discussed them in the pet and dairy sections, and despite their size they are stocky and muscular and produce excellent meat as well.

Cross-breeds as Meat Goats

Some farmers who raise meat goats choose to cross breed Boers with the larger dairy goats such as Nubians, Alpines, Toggenburgs or Saanens.

These cross breeds make great brush goats and tend to be very hardy. Bear in mind that if you are using your large goats as brush goats, they need connection with you and training, preferably before they get too big.

Nutrition for Meat Goats

If you are raising meat goats, you are trying to raise goats with a lot of high-quality muscle. They need a specific and optimal diet. There is a lot of science behind the recommendations for meat goat nutrition. I suggest starting with the article from the North Carolina State Extension which includes the following quote:

> *"...profitable meat goat production can only be achieved by optimizing the use of high-quality forage and browse and the strategic use of expensive concentrate feeds. This can be achieved by developing a year-round forage program allowing for as much grazing as possible throughout the year.*
>
> *Many people still believe that goats eat and do well on low quality feed. Attempting to manage and feed goats with such a belief will not lead to successful goat production.*[76]

Part 3: Maintenance

Feeding, Nutrition, and Treats

This is a general guide to feeding goats. Meat, dairy, and fiber goats require special consideration because their bodies are busy producing. Those particular needs are discussed in their sections.

Because goats will try to nibble on clothes, foam handles on carts, etc., there is a misconception that goats can eat anything. They can't.

Goats are browsers. Unlike grazers (like cows and sheep who eat grass), goats eat woody plants, bushes, trees, shrubs, and weeds. Goats who are put in a grass pasture with grazing being their main food option will be prone to parasites, and their bodies are not adapted for that diet.

Let's look at:
- Feed
- Minerals
- Baking Soda

and
- Treats.

Feed

Chances are, you will need to supplement the browse available to them. Grass hay is the standard choice for this. They can be fed grass hay twice a day without concern. *Alfalfa* hay has more protein (and is more expensive) and can be fed twice a day along with browse.

Some goat farmers like to supplement the grass hay with alfalfa pellets because less is wasted. That's a great strategy but remember that the goats will need the long grass and browse as well – it's not good for their digestive systems to live on pellets, because you will have health issues.

Grains must be monitored closely, as goats are prone to bloat, which can make them very ill (and miserable) or kill them (often quickly and suddenly).

The goat world "rule of thumb" is to only allow a maximum of 1-cup of grain a day per goat. Goats are very motivated to eat grain; they love it. Make sure you keep it secured because they will literally eat themselves to death if they can get into it. Some dairy goat farmers use that 1 cup of grain a day for the daily milking motivator. I like to use grain as a treat and motivator, and sometimes as a nutritional supplement, especially in the winter.

Make sure that you review the list of plants that are toxic to goats in this guide so that you know if there are any plants on your land that you need to avoid. In addition to using that list (or any other that you find), talk to your vet about your specific area. If you contact a local brush goat business, they can also tell you what to look for in your area.

Minerals

In addition, to feed, goats need minerals, particularly copper supplements.

Ask your feed store about loose minerals for goats – it can just be put out for them to access any time. They will self-regulate well. There are feeds with a copper supplement that may be an option for you. In my first small herd of goats, I had a Nubian who was not looking well, and her coat was fading and dry. It turned out she had a copper deficiency. My vet gave me some copper to put in her water every day, and that fixed the problem; she perked up. After that, I have always used feed with copper added in addition to my loose minerals and grass hay.

The minerals are an area where it is good to get local advice. The amount of minerals in the soil that is feeding the goat browse will vary by location. Some will be copper-rich, others copper deficient. The same goes for other minerals they need. Talk to your vet and local goat owners.

Another source of vitamins and minerals is black oil sunflower seed. I like to mix in a handful in their feed every day – you'll get beautiful coats, richer milk, and strong, healthy muscles.

Baking Soda

Put out a block of baking soda for them to access when they want. Again, they will self-regulate very well. Baking soda is an antacid for them and will help prevent bloat. It will help keep their digestive systems healthy.

Treats

As with any treat for any animal (including us humans), goats should not get too many treats. That's why they are treats and not part of their feed.

There are various brand choices for goat treat pellets that are highly effective in getting a goat to come to you. Fruits and vegetables are also a great treat for goats. Whatever human food you give them should be healthy and not processed – goat's digestive systems are not suited to potato chips, etc.

Here is an excellent guide with more detail about the digestive system of goats, how it works, and more about other supplements for goats.[77] Talk to your vet, talk to local goat owners, do some reading, and then decide what the right mix will be for you. As long as you have the basic feed, minerals, and baking soda available to them, you are off to a great start and can choose how you want to fine-tune your goat's nutrition for optimum health.

Shelter

Generally, goats like to be outside and free, but when they need shelter, they *really* need it. Goats do not do well getting wet or in the wind. If they have been brushed out or sheared in the spring and there is a cold snap, this is critical. Even your brush goats will need shelter if you are leaving them in a landscape overnight.

You don't need anything fancy for a goat's shelter. There are lots of resources for building your own low-cost goat shelter that is easy to build.

DIY Goat Shelters

Consider your space carefully before you build your shelter. Make sure you don't put it near a fence, as the goats will figure out a way to get on top and jump out of the paddock. You can also consider whether the wind tends to blow from one particular direction and build your shelter with the back facing that direction.

Here are some of my favorites.

I love the simplicity and sturdiness of this one. If you live in a place with cold winds, you need to add some particle board to keep out draft or line the interior with straw bales. Neither would be difficult or expensive. One thing I really love about this one is that it is *deep* so that if it were cold and even if the wind, rain, or snow was blowing in, the goats can huddle together towards the back. Depending on your climate, you may not need to have any protection in the front as a backup.

You Tube The Tactical Homesteader "Our Goat Shelter Using Free Pallets
http://ow.ly/8vM150CHd8t

If your space allows it, the back of the goat shelter can be up against your house or garage or shed as long as that does not give the goats a way to get out. This insulates well and saves you from having to build anything but 2 walls, a roof and some supports in the middle.

If you live where there is snow, then you will most likely want a sturdy roof that is a "lean to" or "A-Frame" style.

Here is a simple "lean-to" that could be scaled to size if you have more than 2 goats. Note the tips on the straw used to keep them warm as well.

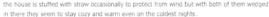

the house is stuffed with straw occasionally to protect from wind but with both of them wedged in there they seem to stay cozy and warm even on the coldest nights.

All that straw makes really good summertime mulch and fertilizer to put on the garden.

Pristine Farm Experience "DIY Goat House"
https://pristinefarmexperience.wordpress.com/tag/diy-goat-house/

This 2-story A-frame adds a little climbing fun and levels to play on for your goats. That bottom area could be for storage or used as another goat shelter space.

Sheepy Hollow
http://ow.ly/MH8w50CHds7

Unless there is a storm blowing wind or rain straight into the shelter, you only need 3 sides. Goats prefer an open space rather than a closed barn-style shelter. In a northern mountain environment where there may be high winds with rain or snow, I like to have a couple of pallets ready with particle board attached to them so that I can stand them up as another wall with space for the goats to be able to get out. For my first goat winter, I attached them with bungee cords on one side so they didn't fall down; then when I figured out that it worked well, I made "doors" out of the pallets so that they can swing open and shut. Never shut your goats in completely unless you are experiencing extreme conditions for a short period of time. Always leave them a way out and plenty of ventilation. Under most conditions, the shelter is open on one side.

You might have an old camper van cover or may be able to get one free from Craigslist. A camper van makes a sturdy roof. Here's a person who used one to make a goat shelter. There are no instructions on the site, but you can see that they were clever using the support in the middle as a space of protection from the elements if necessary.

Summerville Novascotia
http://summerville-novascotia.com/PalletShed/

My own goat shelter is similar to the following "super frugal goat shelter".

Boots and Hooves Homestead "Build a Frugal Goat Shelter"
https://bootsandhooveshomestead.com/build-frugal-goat-shelter/
(There is a material list and step by step instructions.)

Here is another pallet shelter. The doors have hinges and
block the wind from that direction.

The cord you see is to heat a water bowl, not the space. Notice that they have discovered that leaving the pallet spaces open in the back and on the sides is fine. In this case, the wind blows in consistently from the front of the shelter where the doors are. The doors can be closed enough to leave space for the largest goat to get in and out and still be warm enough for them in deep winter.

Goat Shelters to Purchase

If you don't want to make your own shelter, search for woodworkers who make chicken coops or milking sheds for cows. You may also find local people who sell the perfect goat shelter as a "woodshed" that you will see along the roads or highway in your location.

Here is the standard goat shed from one family-owned company. They have a lot of other options as well as you can see from their site. The pricing depends on your location and features that you choose.

KT Barns
https://www.ktbarns.com/goat-sheds.html

If you want something more extensive that includes feeders, here is a company that makes a goat barn and chicken coop combined. If you just wanted a goat barn, they may be able to customize your order, but if you have chickens as well this could be an option for you.

Pa Dutch Builders
www.padutchbuilders.com

Whatever your choice of shelter, remember the basic criteria for your goat shelter is:
- Ventilation but keeps out draft and cold/wet winds
- A solid roof
- Enough room for them to use as shade in the summer as well as curled up tight together in the winter.

Bedding

Goats don't really need bedding as other livestock do. This makes them lower maintenance and easier to clean. If you live place with cold winters, you can use straw for warmth and do a "deep litter" method where you start with some straw and add layers of fresh straw so that the bed is clean, and it forms a compost that is insulating and gives off heat. I have had success with this; in the spring, I clean it out and use it for compost. In the other three seasons, goats are happy with just a clean space out of the wind and rain.

Fencing

Goats are known to be escape artists. They are highly intelligent, observant and will even learn to unlatch gates with their mouths. Adults and older kids will look for ways to jump over a fence (not necessarily because they are unhappy, but just "because it's there"), and just when you think your fencing is secure, your baby goats squeeze themselves through spaces that you did not think was possible.

Fencing has a dual purpose:
- Keeping goats in
- Keeping predators out

To keep goats in, the height of your fence will definitely need to be at least 4 feet tall. Some taller goats who are agile and active (e.g., pack goats) may need additional height. Goats will often rub against the fence and gates, so it needs to be solid and sturdy. Earlier, we discussed temporary electric fencing for brush goats, but that is not recommended as a permanent solution for your home paddock.

Here are 2 close-ups of the extra fencing we installed in places around gates. The kids were getting out of these places; now, it is secure.

See the extra wire fencing that enables the gate to open but keeps goats inside. (The livestock guardian dog noticed my attention on the gate, so he had to get in on the act.) When you open the gate on the right, the fencing that covers the wood post opens with it.

To keep predators out, you need to assess who the predators are. Domestic dogs cause most goat deaths from predators. If you need to keep dogs out, you may need a fence higher than 4 feet because many dogs can easily jump over a 4-foot fence. If you are in a rural area, you may also have coyotes, mountain lions, and bears. Predation is another area where local knowledge of the local wildlife is critical, in addition to knowing your threat of domestic dogs.

These instructions are the best guide I could find to take you through the steps of building your fence. It talks you through the details of the actual materials and building and the considerations for your specific landscape. Use this guide if you need to build your fence or adapt it as you incorporate goats into your homestead.[78]

Basic Goat Health and Wellness Maintenance

Goats are generally hardy and easy to care for, but there are some essential basics. As with all animals, "prevention is the best cure". We've discussed shelter, fencing, and nutrition. Here are other basic needs that are critical:

Water Water Water Water

Every mammal needs water, goats require a lot less than cows or horses, but they still need adequate access. The particular issue with goats and water is that they are picky. They will stop drinking and dehydrate themselves unless it's fresh and clean. They will not drink dirty water.

At the very least, use a water container that is heavy and solid enough not to be knocked over, but make sure that you have a way to reasonably turn it over and clean it out. Automatic waterers are well worth the investment. Automation will prevent you from having to dump water out, clean the container, and fill again. It saves water, time, and heavy labor every single day. Automatic waterers are especially useful in cold winters; see the winter section for an example of heated waterers.

Amazon "Animal Hydration Tough Guy Automatic Waterer"
http://ow.ly/aQ5d50CHfsD

Hoof trimming every 4-6 weeks

If hooves are not trimmed, your goats can contract a bacterial infection called "hoof rot". This will cause the goat to experience discomfort and suffer eventual lameness. Also, the hooves grow like our nails. They will literally grow over and around the area where the hoof meets the ground and cause the animal to limp. Hoof rot is of special concern during wet/muddy seasons when there will be more bacterial activity. Besides keeping your paddock area clean, hoof trimming is critical. Even in the dry times of the year, hooves need trimming every 4-6 weeks.

This goat-keeper offers step by step instructions about hoof trimming

Timbercreek Farmer
https://timbercreekfarmer.com/goat-care-and-maintainance/

More hoof trimming instructions and illustrations can be found here[79]

Wellness checks

Every day I give a quick check over my goats, making sure there is no limping, diarrhea, wheezing, coughing, or discharge from their nose or eyes. I also spend enough time handling and petting them to observe their behavior and energy. When I feed them, I watch to see if there is any pattern of loss of appetite. Is one goat uninterested? If any of these symptoms are present, contact your vet. I recently had a kid who was looking tired. We had a cold snap and adjusted with a little more grain – the kid perked up and was soon kicking up his heels and butting his brother as he should.

Count your goats

I count my goats twice a day. We live in a wild place, and even though there is a livestock guardian dog, I make sure that nobody has escaped or been snatched. Remember, most goat deaths are from domestic dogs. If you live in an area with dogs around, you are statistically more at risk of losing goats to predation.

Vaccines

Discuss this with your vet; typically, goats are vaccinated for Tetanus and CTD (clostridium). Either one month before or after those vaccines, they are also vaccinated for rabies. When you purchase a goat, get a printout of the vaccines the goat has had from their vet. If you sell a goat, make sure you have printed out (or can email a .pdf) of the vaccines the goat has had.

Also, follow your vet's advice about wellness checks and regular visits.

Handling and Interaction

Whatever the function of your goats (pets, dairy, meat, fiber, or working), they need to be used to being caught and handled. You need to examine them, treat them if they get a cut or need other care, trim their hooves, and work with them for whatever function they fulfill.

When you get new goats, then get them used to being handled by petting, scratching, touching their bellies and legs, and picking up their feet. When you first get them, do this once or twice a day. When they are entirely comfortable with this, then just keep it up every couple of days and spread it out to a couple of times a week.

Now that I have more than five goats, it's too much to do this to all of them at once, so after the newcomer has been handled every day until they show they are tame and docile, I rotate them through the herd, giving attention to a couple of goats every day.

Toys, Games, and Interest

Any animal is unhappy when they are bored. Goats are active and intelligent. In addition to needing connection and handling, they need some mental stimulation. Like any animal (including humans), a bored goat is a goat who at the least is unhappy, and at the worst, may start destructive or aggressive behavior to act out.

The agility activities we discussed earlier are ideal, but you may not have the time or inclination to train your goat for those activities. Even if you do have time for agility, the goats need some things to play with inside their paddock when you are not there.

Things to Jump Up On

If an urban person knows anything about goats, they know that they like to be the "king of the hill". Goats are famous for jumping up on top of things because that's one of the things they love to do most.

Give them some fun options; here are some ideas for DIY and reuse:

Tires are versatile and are always a big hit. You may have a set of tires you can use creatively, but if you don't, you can probably find someone who would be happy to give tires away. This photo used tractor tires so that the tires serve as both tunnels and things to jump on, but smaller tires work as well.

These tires are stabilized by burying about 1/3 of the tire in the ground to hold it up.

Boer Goat Profits Guide
https://www.boergoatprofitsguide.com/goat-toys/

If you have hard or rocky soil and digging is not an option, then stacking tires makes a great playground as well. Just make sure they are stacked so they won't fall when the goats are jumping on them.

This dual-purpose tire stack is genius, the creator said that they poured cement into the tires and it helps keep their hooves dry and trimmed. Apparently it's a good day bed as well.

Pinterest – no instructions but easy to copy the idea
https://www.pinterest.com/pin/425519864765709464/

Here's a simple box that this billy goat is currently owning. One of these, a couple of stumps and a couple of tires can provide options for your goats to play.

One of the kings next door

If you have the space to put something this high well away from a fence, a tiered playground can be constructed with reclaimed pallets and wood

Boer Goat Profits Guide
https://www.boergoatprofitsguide.com/goat-toys/

In the agility section we looked at stumps – gather one, two or a few or a lot! If there are some that are large enough for them to lie on, your goats will enjoy lying down on a raised space. The smaller stumps will provide the jumping factor as well. Another beauty of stumps is that they are easily moved around to make variety and change it up.

Goat Swings

If you have the building skills, this is simple, and it looks like the goats are really happy with it.

Pinterest – no instructions – but such a great idea!
https://www.pinterest.com/pin/477874210465906698/

Goats like to butt. They can have a lot of fun with a Theraball or a used 5-gallon water jug. Anything that rolls well!

Boer Goat Profits Guide
https://www.boergoatprofitsguide.com/goat-toys/

Yoga balls are my favorite choice for goats pushing things because they seem to love them and are incredibly entertaining.

I have bookmarked this video to watch as a stress buster – goat vs ball – I'm not sure who wins, but it makes me smile and laugh a lot!

YouTube – Mary Ann Johnstone "Best use for a Yoga Ball, According to my Goats"
https://www.youtube.com/watch?v=mu31o9zzHQs

Mini trampolines

People sell (or even give away) trampolines every day – goats love them!

These little babies don't have the jumping down yet but notice that they set up stumps and a board to make a bridge next to it. It won't be long before they realize they can jump onto the trampoline from the bridge!

YouTube – Kul Farm – "My Goats Reacting to a Trampoline"
https://www.youtube.com/watch?v=TvAwgYlCLRQ

These kids have had more experience and have learned that jumping is fun.

YouTube "Bouncing Kid Goats on the Trampoline"
https://www.youtube.com/watch?v=NQ-2KtJ8hOo

Tie a stiff push-broom brush to a pole for scratching

Pinterest – anyone could do this!
https://www.pinterest.com/pin/277252920797685037/

Make a teeter totter!

Here's a video of a bunch of different teeter totters for goats. Some are super simple like this one in the preview below, others are more complex for fun and more challenge. It's very easy to make something that will be entertaining for both you and your goats.

YouTube – Crazy Funny Stuff CFS – "Goats on Teeter Totters"
https://www.youtube.com/watch?v=OZoP_9ae8Q8

Winter Care

Most goats are adaptable to cold winters, and quite a few breeds originated and then developed in northern places at high elevations with harsh winters. For most breeds, a little care and attention over the winter will enable them to get through the winter just fine.

Winter Shelter

We discussed shelters above, make or choose a shelter that has doors, and then leave them open a little bit so that the goats can get in and out.

If you want to make it extra nice for your goats, line your shelter with *straw* bales (*not* hay bales, they will eat them). Straw is a soft, warm insulator and will block draft and wind while allowing some ventilation. Just make sure that your goats have proper ventilation. It is dangerous for their respiratory health if the shelter is not adequately ventilated, especially since they may be spending hours a day inside.

A trailer can serve as an extra "living room" in the winter. It can give the goats a different place to be and break off into smaller social groups if they want. It also reinforces the trailer as a comfortable and safe place, which is great if you use it to transport your goats.

Winter Bedding

There needs to be some warmth in harsh winters, and straw is the best for this. Since shoveling out the bedding every day in the winter is very difficult, we suggest the "deep litter" method for the winter. Add a layer of fresh straw every day so that they are lying in clean bedding; the lower layers will begin to compost. This has the added value of heat generation over time. In the spring, you have the once-a-year project of cleaning it all out and composting it, but many goat keepers agree that a deep litter is much less work than cleaning out the shelter every day in storms with accumulated snow.

What about heaters?

Goats don't need heaters, and they are a fire hazard. Also, their coats will not adapt to the colder temperatures, then if there is a power outage, they would suddenly be plunged into temperatures that would be highly dangerous to them. Let the shelter itself, deep litter, and straw bales (if you choose) serve as their warmth; their winter coats will take care of the rest.

Boards and Bridges

When it's not storming, your goats like to be outside, but they hate getting their feet wet. Adapt your paddock to include more bridges to help them move around without their feet being in the wet so much. Deep snow is both wet and very cold and uncomfortable to goats.

This is critical during mud season. As the snow melts when the temperatures warm, bacteria thrive. Layout boards on the ground and figure out ways to help your goats keep their feet out of the mud. They will love you for it, and you'll save yourself from hoof rot risk. It is recommended that you check the hooves for trimming every 4 weeks when it is muddy.

Heated Waterers

If you live in a cold climate, you will have to invest in heaters for waterers. Here are some examples:

It is easy to find heated buckets like this

Amazon "Gallon Watt heated Bucket"
http://ow.ly/tP0k50CHgew

If you want to continue to have an automatic waterer during the winter, you can check out the eight automatic heated waterers at the bottom of this page:

PeteCo Supply Co. Omni-2 Special #18680
https://www.waterls.com/SheepFountains.htm

Two important notes about winter water:

- Water should not only not be frozen; it should not be too cold either. Drinking cold water will lower the goat's body heat, and they will dehydrate themselves. Keep the water clean and cool but not cold.
- We think of animals needing more water in heat, but cold winters also require more water for different reasons. Wind is dehydrating, and the goats are likely to have more dry feed such as dry hay and pellets than fresh browse that contains more water naturally.

Make sure your goats have enough fresh water at a drinkable temperature.

Goat Jackets

Goat jackets serve three purposes:
1) Keeping the goat dry. This is really the main purpose. Whether snow or rain, your goats need to be kept as dry as possible
2) Warmth
3) They are very useful to keep a goat from scratching or biting at a sore or other irritant.

Valley Vet "Goat Coat with Neck"
http://ow.ly/m8vp50CHgtj

I like this one above because it includes the neck which helps keep water from dripping down when it is raining. It also isn't too heavy and hot. Here are more resources for various kinds of goat jackets, including thicker fleece for warmth.[80]

If you use goat jackets, take care not to leave them on all the time as it makes an optimal environment for parasites.

Feed in Cold Winters

Any mammal burns more energy in cold weather just keeping the body warm. Increase your goat's feed in the winter. When I moved to a cold climate, I asked my vet about this and she suggested increasing the goat's feed 1% for every degree below 32 degrees F. This is documented as a general suggestion by professionals.[81]

Besides increasing the amount of feed, increasing protein is also recommended. Mix in more alfalfa hay or add in some alfalfa pellets.[82]

Follow the guidelines we have discussed, and you can prevent winter health problems for your goats, and a lot of work for yourself.

Possible Goat Ailments

You need to know about the possible conditions, ailments, and diseases that threaten the health of your goats. Do not let this list make you think that it's difficult to have goats. As discussed earlier, they are hardier than other livestock such as sheep and horses. Most of these are preventable through either a vaccine, deworming, or your goat's environment protocols.

We have discussed the diseases transmittable to humans in the FAQ:
- Chlamydiosis
- Q Fever
- Leptospirosis
- Ringworm
- Salmonella

This is not intended to be an exhaustive list. It is a curated list based upon the most common ailments you need to watch for. Ask the advice of your vet to learn about others as well.

Caprine Arthritis-Encephalitis Virus

Caprine Arthritis-Encephalitis Virus (aka CAE or CAEV) is usually transmitted from mother to kid at birth through feeding or transmitted through feces or blood to other goats. There is no cure; the only way to stop it is to ensure that a goat you have purchased has tested negative and to test your goats regularly. (Ask your vet about time interval recommendations for your situation.) CAE does not transmit to humans but can take down an entire herd.

The Cooperative Goat Extension lists these symptoms of CAE: [83]

Arthritic CAE
- Lameness (might be sudden)
- Stiffness
- Reluctance to walk
- Abnormal posture
- Reluctance to rise
- Weight loss

- Swollen joints
- Walking on knees

Encephalitic CAE
- Incoordination
- Inappropriate placement of limbs
- Progressive paralysis
- Depression
- Blindness
- Head tilt
- Seizures
- Death

Pneumonic CAE
- Deep, chronic cough
- Difficulty breathing
- Weight loss

"CAE virus may also cause a chronic wasting disease in which goats continue to lose weight although appetite is unaffected." [84]

Coccidiosis

Coccidia is the most common parasite found in goats. When they multiply exponentially, it is called coccidiosis. Coccidiosis can be prevented as it is contracted through fecal matter going to the mouth. Coccidiosis is one of the many reasons you want to keep a clean paddock and why grazing is not the best for goats as grass is closer to the ground than the brush that their systems handel well. Young goats are more prone to coccidiosis because their immune systems have not developed enough.

Symptoms
- Loss of Appetite
- Diarrhea
- Dehydration
- Fever
- Weight Loss or Poor Growth
- Hunched Appearance

Treatment

- Fortunately, coccidiosis is easy to cure. You must get a prescription from your vet specifically for coccidiosis.

<u>Prevention</u>
- Reduce the amount of exposure to contamination by keeping the bedding, paddock clean, food, and water sources clean. Don't forget the unofficial "hang out" spaces where the goats might lie down in their own poop.

Here is more information from Michigan State University Extension about the importance of the prevention of coccidiosis. They also provide detailed timelines and descriptions of how the infestation develops. [85]

Wether Stone Formation [86]

Wether stone formation results in urethral blockage. In humans, this is known as "kidney stones". When stones form and become larger, they make their way to the urethra, causing the urine to back up and potentially rupture the urethra or bladder.

Wethers are most vulnerable to this condition because sometimes the urethra is narrowed when they are castrated at a very young age.

<u>Symptoms</u>
- If a male goat (usually a wether) is suddenly lying down and crying out in pain, this is one possible cause.
- Prior to being in full-on agony, the goat may have acted a bit "off" like he is constipated or not eating as much.

<u>Treatment</u>
- If it gets to the point where there is a blockage, euthanasia, or emergency surgery are the options. Prevention is the key to this condition.

<u>Prevention</u>
- Proper nutrition. Don't let your goats have too much grain. A diet with a lot of grain raises the PH in the bladder, and stones begin to form.
- Measure and track the PH of your wethers. The normal range is 5.5-6.5. If you see it creeping up higher, you can adjust their diet.
- Watch your goat's weight. Obesity makes them more vulnerable to forming stones.
- Ensure your goats are hydrated.
- Make browse readily available. Diets high in alfalfa hay and clover can encourage the growth of stones.

- Wait till your male kids are 3-4 months old before castrating them. If you acquire a wether, ask what age he was when he was castrated. Allowing them to develop a bit longer enables the urethra to grow to full diameter.

Hoof Rot

Hoof rot is a contagious bacterial infection. It is one of the main reasons to keep a clean paddock. The bacteria that cause hoof rot lives in the feces of goats and sheep and thrive in wet conditions.

Symptoms
- It begins with a sore in between the toes where the bacteria have started to grow.
- As it progresses, the area between the hoof's sole and outer wall will begin to break down.
- The outer part of the hoof may break away from the inner portion when trimming
- A rotting smell from the hoof
- Limping
- When hoof rot becomes severe, the goat may be in so much pain that they will graze on their knees.

Treatment
- If you are catching it in the early stages, you can trim away the infected area. (Make sure you get all of it.) If you are trimming your goat's hooves every 4-6 weeks (every 4 in wet seasons), then you will most likely be able to get control and be able to trim it out yourself. If not, then get your goat to the vet. Hoof rot is not only affecting the health of that goat but is also contagious to the rest of your herd.
- When you trim, apply "Hoof 'N Heal"[87]
- If the infection is not eliminated, or if it seems to have spread to multiple goats, then contact your vet to help you deal with it.

Prevention
- Trim your goat's hooves every 4-6 weeks. Every 4 weeks in wet conditions.
- During muddy/wet seasons, give your goats planks to walk on from their shelter to their food and around the paddock. They hate getting their feet wet and will gladly use the planks.
- Keep your paddock clean.

Bloat

Bloat is simple but can be deadly. Goats are not as prone to bloat as sheep, cows, and horses, but it can happen.   It is preventable.

There are two types of bloat: [88]
1. Frothy Bloat
    - It occurs when the rumen is blocked so air cannot escape. As a result, the air builds up inside the rumen.
    - Frothy bloat can be caused by the goat eating too many "lush legume" plants such as green feeds, new hay, clover, alfalfa, or wet grass. Another cause is "sudden access" to a lot of grain.

2. Free gas bloat
o It occurs when the esophagus is blocked, and air is trapped in the upper part of the rumen. This is usually from too much grain, apples, or carrots.

Symptoms
- The goat will not eat normally, will be restless, and generally look uncomfortable.
- The left side will begin to bulge out.
- Biting or kicking at the left side
- Collapse

Treatment

If you think you want to self-treat your goats in the event of bloat, then talk to your vet, purchase the right supplies, and be ready before it happens. You don't have time to learn when there are symptoms; you have to act.

"Treatment includes careful passage of a stomach tube; this should be curative in the case of free gas bloat. If the obstruction cannot be corrected with a stomach tube and free gas bloat continues to develop and threaten the goat's life, a veterinarian may need to trocharize the rumen. For frothy bloat, drenching with poloxalene or mineral oil (100-200 cc) may help. **DO NOT** *drench mineral oil without a stomach tube, or it will end up in the lungs. Walking the goat and massaging the flank may be of value. Determine the cause of the frothy bloat and address it."* [89]

https://goats.extension.org/goat-bloat/

Prevention

Fortunately, bloat is preventable.

- Ensure that your goats do not get turned out into a green, wet lush pasture, and consume a lot.
- Make sure they don't get too much grain.
- Watch the amounts of apples and carrots – don't give them a lot at once.
- Don't give them too much new hay, clover, or alfalfa.

External Parasites[90]

External parasites that will attach to your goats include lice, fleas, ticks, mites, and nasal botflies.

Symptoms

Lice

- Lice are the most common external parasites. There will be hair loss, itching, rubbing up against surfaces.

Fleas

- There is usually no hair loss that results from lice, but your goat will be very itchy. You and your dogs can get them too and then bring them into your house.

Mites

- Mites can be on the skin or in the ear canal.
- If on the skin, the goat's coat will be dull and fall out in clumps.
- If in the ear, the goat will shake its head and scratch at it.

Ticks

- The best way to tick check your goat is with a fine comb, combing the coat backward. Look for any that are loose and watch for any that have attached.
- There are often no symptoms until the tick is full of blood. At that point, the goat will try to scratch or bite at it as it will be causing discomfort. You will find the tick attached.

Nasal Botflies

- These botflies climb up into the nasal passage and lay their eggs. When the eggs hatch and the botflies grow, the goat gets a lot of discharge from the nose and will shake its head, blow, and snort to clear its sinuses.

Treatment

Lice
- You may need to shave the goat, but my first line of defense is to (carefully) apply Diatomaceous Earth (DE). Make sure that you do it outside and that you or your goats do not breathe it. DE is the skeletal fossil remains of tiny creatures whose sharp edges will tear apart the parasites. It is very effective but will also tear apart lungs. Apply it every 2-3 days until the lice are gone.

Fleas
- Shaving is best to eradicate fleas. Apply DE Fleas love to hide in spaces, so you've got to clean out your paddock and barn thoroughly – including any cloth. Make your goats stay outside, put on a dust mask, and apply a coat of DE to their barn or shelters.

Mites
- For the mites on the skin, shave the goat and apply tea tree oil with a spray bottle or a cotton ball. (Never full strength, follow directions for dilution). That kills them. Some people use DE for mites on the skin as well.
- For the mites in the ear canal, you can inject olive oil into the ear. Consult your vet before you do this to get recommendations for how much and how often. They will recommend a syringe that will work for this purpose. Do NOT ever put DE in an ear canal.

Ticks
- Remove the ticks with tweezers – try to get the whole head, don't just pull on the body. Put the tick in 70% rubbing alcohol or make sure it's torn to pieces. If you just drop it, it is likely to attach again.

Nasal Botflies
- Consult your vet about a prescription level de-wormer that will work.

Prevention
- Keep your paddock and shelters clean
- If you live in an area with many of the parasites we've discussed, you might apply DE to the shelters proactively.
- Get some backyard chickens! They are amazing with pest control.
- For the flies, you can set out flytraps in the trees in the summer.
- Don't leave a jacket on a goat for extended periods. That gives the parasites an optimal environment. Use jackets for keeping dry in the rain, for the bitter cold when they need to be outside, or as a temporary barrier to an itchy place or wound. If they are itchy because of parasites – don't use a jacket!

Caseous Lymphadenitis (aka CL) [91]

Caseous Lymphadenitis is a contagious bacterial infection of the lymph system that causes abscesses. The abscesses may be internal or external. It is transmitted through direct contact with the pus from one of the abscesses or through contaminated soil.[92]

Symptoms

External CL usually exhibits abscesses behind the ears, around the neck, but sometimes between the legs and around the udder or genitals. It is a condition of the lymph system, so the abscesses tend to be around lymph nodes. The animal is not extremely contagious until the abscess breaks, and the pus runs out. At that point, it is highly infectious and can be transmitted through direct contact or with surfaces or soil that is soiled with the pus.

Internal CL is less common in goats; it is not until the goat is dead that the abscesses can be examined.

Treatment

There is no treatment for Caseous Lymphadenitis. The animal must be put down to avoid spreading the infection to the entire herd.

If you see an abscess, especially one around the neck, jaw, genitals, or where the legs meet the torso, contact your vet and get it tested.

Prevention
- Get the vaccine. This disease can be prevented.
- Have a closed herd policy. Ensure that any new goats have tested negative and don't allow other goats or untested livestock to contact your goats.

Pneumonia [93]

Pneumonia is an infection in the lungs and can be caused by a variety of microbes and causes. Inadequate ventilation can contribute as well as overcrowding.

Symptoms
- Dull, lethargic, uninterested in food.
- Fever, cough, difficulty breathing, runny nose.

- Pneumonia requires a vet's expertise as the treatment will depend on the bacteria causing the illness. You can also consult with them about any environmental contributors such as poor ventilation in shelters.

Prevention
- Isolate the sick animal to avoid transmission to your herd.
- Provide proper ventilation and enough space for your goats.

Introducing Your Dogs to New Goats

Dogs can harass, injure, and kill goats. As we said in the beginning of this guide, domestic dogs are the number one killer of goats every year. Even a small dog can panic and chase goats to exhaustion. If you have a beloved family dog, you want to set yourself up for success when the goats are introduced.

There are numerous stories of tragic suffering and losses of goats caused by domestic dogs, and if a dog kills a goat it is strongly suggested that they must be put down.

It is understandable that goats are at least wary if not frightened by dogs. Your goats may have been around dogs who were good with goats, they may have never seen a dog before or they may have had bad experiences with dogs. If your new goats have been harassed by dogs in the past, then they may never really "get over it", even if your dog is trained well and does not bark or chase them.

There are numerous methods for training dogs. Your own skill, experience and choices factor in to how you will train your dog to be safe with your goats, as well as the dog's breed and individual temperament.
Choose your own commands and training method; we recommend that you consult a professional trainer who has experience with dogs and goats. Here are some basic principles for training.

Contain your dog inside when you arrive with your new goats to get them set up. Don't allow them to run up to the vehicle barking. Even if your dog is curious and sweet, it will be overwhelming and stressful to your goats when they feel very vulnerable. Protect your goats and let them be introduced to the paddock in peace.

Basic Training and Introductions

1) Bring your dog outside the fence on a leash a few times day. I would make mine sit quietly and watch the goats. If a goat is curious, comes to say hello and sniffs noses, that's great. This can be the beginning of normalizing with the dog. Do it a couple of times a day for 5-7 days. Watch the body language of the dog and the goats. If the dog has a slow, flat tail wag, and gently sticks their nose through the fence, that is a good sign. If the tail is up and stiff, the dog is not relaxed, and you need to be ready to take action. If your dog lip curls or growls or barks get them out of there immediately. The goats need to know that you will protect them.

2) If the dog wants to bark or acts aggressive at the fence, then stop the process and get professional advice from a local trainer with experience. You don't want to program your dog for this behavior, and you don't want your goats to be afraid (or potentially aggressive) to your dog.

3) After 5-7 days of the supervised, leashed greetings outside of the fence, assess whether your dog is ready for the next step. If your dog is relaxed and will sit and watch the goats, wags and is friendly if they approach, you have been successful. There may even be particular goats who are starting to make friends with the dog, but that is not necessary to continue. Success is your dog being relaxed and adapted to the goats, and vice versa. If that is the case, then you can lead them inside the fence *still on a leash*.

4) When you bring them inside the paddock, make the progression slow and incremental. Bring the dog in, make them sit. If the goats skitter or are wary, you might decide to repeat that step: go inside the gate and stop. Do that for a few sessions until the goats have adjusted. When the goats are relaxed, they are likely to come up to the dog. If that is the case, let the dog stand and greet the goats. Be ready to pull them up to a sit if necessary.

You decide what words to use for commands that are consistent with your training. Some goat keepers recommend using treats for both goats and dogs in this process, I have found this to be too much of a handful holding the leash, paying attention and giving treats to excited animals. I prefer to reinforce good social behavior with my voice, lots of praise, calm and assurance. If your dog is clicker trained, you might incorporate that. Use whatever commands and methods are consistent with the training that your dog is accustomed to.

5) Once the dog can come inside the paddock on the leash and be calm and well behaved, then they are ready to be led around while you do some chores. If possible, it helps to

have a second person for hands. If you don't, then tying the dog to a fence might also work, as long as they don't become agitated when you walk away from them to do things.

Decide when it is safe to allow your dog off leash while still supervised. Many goat keepers recommend never allowing a dog to be unsupervised in a goat paddock unless they are trained livestock guardian dogs. There are stories of people who thought their dog was fine with the goats and then suddenly the dog went into predator/chase mode with tragic outcomes.

This is my basic protocol, there are many ways to introduce your dog to your goats. This footnote contains resources that provide other goat owner's insights and methods.[94]

Many goats live happily with dogs. For many goat keepers all it takes is the right introduction to train the dog and "wire" good behavior. Be attentive and consistent, err on the side of caution and be ready to seek professional advice if needed.

Livestock Guardian Dogs

Livestock guardian dogs (aka LGDs) are amazing creatures. There are a number of traditional breeds, the thing they have in common is that they were bred to be able to bond to a herd, be incredibly protective and be solitary. They were originally bred to be able to think for themselves and stay with stock for days on their own. They are extra-large dogs, who are fierce with predators or intruders.

LGDs are working dogs, not pets. They can be trained to be safe, friendly, social and even loving to children and other family members, but their job is to be with the stock around the clock.

One of our current LGDs is a big fluffy "snugglewump" who has bonded with the goats. The youngest 2 kids started curling up against him when they were about 2 weeks old.

Here are some resources if you are interested in getting an LGD.[95] We've had three of them, here are some things we've learned:

Livestock guardian dogs can be aggressive to other dogs and humans. We have trained all of our LGDs to be social, but we have noticed that online forums and groups include people who let their dogs be out with their stock without much interaction and have had to find other homes for their dogs due to aggression towards humans and neighborhood dogs.

LGDs will have a strong sense of wanting to walk *what they believe* is their perimeter. This may not be the same as *what you and your neighbors see as* the dog's perimeter. If an LGD gets out, they may be aggressive or even attack the dog or human next door. That "snugglewump" I mentioned earlier is so gentle and loving to humans he knows and the goats but cannot be allowed to escape as he can turn into 120 lbs of fierce attitude, muscle and teeth with other dogs. This has been true even with training, we continue to work on it and it gets better.

Great Pyrenees have long memories. Our neighbor's dog stole one chicken and the LGD would chase her with teeth barred even 2 years later. They don't forget. Prior to the chicken incident they were friends and playmates.

LGDs bark. Colorado Mountain Dogs are a recent breed developed to be more selective in their barking (along with other desirable traits).[96] Most breeds are not selective in their barking. A lot of the deterrent to predators lies in making their presence known and voicing threats. Some inquirers ask whether you can train an LGD not to bark or use an anti-barking collar on them. One author describes this as "installing a security light with no light bulb".[97]

Extra-large dogs have extra-large appetites.

We've not lost any goats to predators (coyotes, mountain lions, and bears are all here, they are deterred from the goat paddock).

Livestock guardian dogs can be an incredible asset to a homestead. If you decide to get an LGD, be prepared to train them to be social, safe with children, and manage the introduction of any guests or workers you might have coming into the space. Train the dog to accept anyone you have introduced. Join online forums and groups and do research before you get them. Be in touch with someone who has successfully trained them, pay for some training.[98]

Training Your Goats

We've discussed specific training for goats with various functions such as working goats, dairy goats and brush goats. We've also discussed agility and training your goats using games.

There are general principles that go for *any* goat, and training is part of your bonding with them.

As discussed earlier, whenever you purchase a goat, see if you can walk up to it and handle it. Can you examine their hooves? Are they docile when you examine their body for any sores, cuts, or parasites? If not, and you still really want that goat, then know that you are in for some training *right away*.

If you breed your goats or purchase kids, make sure those kids are handled lovingly, gently, and confidently, from a very young age. Build that rapport and trust immediately.

Goats are prey animals, so it is critical that you (the goat leader) are calm and confident. This demeanor will help gain their trust and want to follow your lead. You also need to be assertive because goats are very smart, and they want to establish who is the lead goat. *You* are the lead goat, and they need to know it.

Hitting a goat is not an effective punishment. Ethics aside, at the very least this teaches the goat the wrong things and will establish you as untrustworthy. If they will not trust you, they are less likely to be compliant. They will copy your behavior so they will learn to be aggressive with you and each other. I've known exceptions, but for the most part, most goat keepers agree on this.

The tools for correcting a goat are:
- Your voice
- Your hand moving their head to the side, or putting pressure on the side of their body
- Squirt bottle
- Something thrown at them to get their attention and make them uncomfortable (not to hurt them)
- A long stick to extend your arm and correct direction or make them get away from something
- Interestingly, hitting a goat plays differently in their minds than using a long stick (to extend your reach) to correct the goat, or throwing something from a distance to get their attention. If you need to get them to stop doing something at a distance, don't throw something that would hurt them, but something that will surprise and interrupt them and get them to pay attention to you.

Learning "no"

Goats learn "no". You can shout at them sharply when they are doing something you don't want them to do but save the sharp shouting for when it's important so that it isn't background noise that they become used to.

For example, you walk into the barnyard and there is a goat who will literally eat the clothes off of you if you let her.

- "no" (said firmly but quietly)
- She does it again
- "no" (said a little more strongly plus push her nose away from the side of her neck/head, not her forehead)
- She does it again
- "NO!" "GET!" Shoo her away, stomp your foot

- Go back to being calm, confident, and loving. Expect the need for repetition. You may decide that after a couple of times that this goat needs the strongest "NO!" the first time she nibbles on you but give her a chance to learn first.

My neighbors have a billy goat who loves to eat clothes. He tried it with me, I did this routine once and he never tried it again. They learn, and some learn very quickly.

The rest of the herd will watch and learn as well. It's not just this goat's behavior that will be influenced, *and* you will be reinforcing who you are to their herd. You are the boss. You are the lead goat.

Play with your goats a lot, but *don't play butting games*. This can seem tempting, cute and fun when they are kids, but even a Nigerian Dwarf adult is a formidable force of strength when they decide to butt as an adult. Don't start. [99]

Also, when you correct them physically, don't push on the forehead of the goat (for example to correct the one that eats your clothes). This signals that you are butting them. You don't want them to butt you, so don't teach them to do it. If you need to move their head, push them on the side of the face or the side of the top of the neck, but not straight on their forehead. You must not teach them to butt you.

Training with a Squirt Bottle

The squirt bottle is your friend. Goats hate water. Have several small bottles around so that they are handy and train your goats to associate it with "no!". I put my squirt bottle in a water bottle carrier belt so it's right there and handy. Make the squirt immediately after or during saying the word "no!". [100]

Leash Training a Goat

Even if you are not raising goats who pull, pack, or need to be led regularly, training your goat to be led is very useful for health examinations and the possible need to get them to a vet. It is also bonding and can be a source of fun and interaction with them if you train them well. If you are training for shows or other functions, leash training is critical.

Here is how I was taught, and it worked like a charm:

Bring out a collar, let them sniff it. (It's always good to let a goat sniff something you are putting on them or using on them, it helps them feel safe.) Then put the collar on them, let them get used to it. Eventually I may graduate to a bridle, but it's ok to start with a collar. If you are going to lead them often, then definitely use a bridle instead of a collar as a collar pulls on their neck a lot. You have more control with a bridle as well.

Get them near a fence or post where you have the leash attached, give them a little grain or treats and clip them on.

Walk away and do chores, hang out with other goats, come back, and pet the leashed goat when it's being good, but make them learn that they will get no-where pulling on that leash. I was lucky to get this advice for my first goats and from what I hear from other goat owners, it saved me a lot of time and effort.

When you've done that routine enough for them to learn that they get tied up while you do chores and that resistance is futile, then it's time to start to lead them. It works like a charm!

After they learn that pulling is futile, start teaching commands as a game with treats. Get their interest and pique their curiosity.

- "Go" or "Walk" or "chik chik "lets go!"
- I use "chik chik" generally for "hey! Pay attention now! I'm going to tell you something that might be fun!" Sometimes I'm telling them that I'm here with treats, sometimes I'm telling them it's time to play, sometimes I'm telling them to get into the trailer, or to walk with me. If I need to tell them to get into another area of the paddock, it really helps to not have to herd them through a gate.
- Clicking sounds for "go faster"
- Left
- Right
- Whoa or Stop
- Back up
- Stand

You can also work with any other specific commands you might need in a show ring or for other purposes like getting into a milk stanchion. I have added a command that is similar to "no" but is communicating that I don't like what you're doing, do this, watch me". The command is "tssst tssst!" – listen up! Whatever your commands are, be consistent, teach them incrementally and practice them. Build up the training gradually so that all commands are used.

If you are using a collar, lead them from the top, not pulling on them from the front. With a halter you are leading from the front. If it's goat against human, the goat will win if you are trying to pull from the front.

We discussed collar training in agility goats, but here is the photo of the proper way to lead a goat by the collar. For leash training, obviously you have a leash as well as the collar to hold on to.

The Thrifty Homesteader – A reminder of how to lead a goat
https://thriftyhomesteader.com/working-goats-your-journey-begins-here/

When you lead the goat, and it stops, stay calm and assertive. Just continue putting pressure on the collar. This is a critical moment. If you stop and try to negotiate or plead with the goat, you teach them that they are in charge. If the goat does not go forward with steady pressure on the collar, then say "no!" (– squirt –) and immediately give a command. Lead them right or left or around to go the other direction and then give lots of praise. When your goat balks, be consistent in your response.

Train your Goat to Load and Ride in a Trailer

If you need to take your goats to shows, the backcountry, a brush goat job, or even the vet, you'll need to train them to load and ride in a trailer. If you don't have a trailer, you'll at least need to train them to load up into the back seat of a car to get to the vet when you need to.

Make it a game. If it is possible to get the trailer into the paddock so they can get used to it, that's ideal. If you can't, then get your goats outside the trailer, open it up, jump into it yourself, saying "hup!" or "load up!" run to the back (they will follow you) and give them treats and praise. Run out again "all out!" or "ok!" and provide them with praise when they've come out. Do this literally 20-25 times.

Do it at least twice a week, more if possible. In a couple of weeks, they will have it down.

When your goats are comfortable getting in and out of the trailer, the next step is getting them comfortable with riding inside of it. If you have a partner who is willing to help, get them to very slowly drive the trailer while you walk alongside the trailer with the goats. When they are used to this, then walk alongside them for a short distance (like 1/8 mile), stop and open up the trailer, get in with them.

When inside the trailer, be calm and assuring, a few treats and lots of touch and affection. Ride back to where you started and let them out. When you get out, make it a celebration of praise and treats.

Next time make them take the little trip alone and start extending the length of the journey.

Training is part of interaction and relationship building as well as being critical for practical reasons. When you play games with your goats, that training boosts your relationship and rapport with the goats, which builds your reputation as the lead goat, and they will happily learn other more necessary things from you. Do training that is both practical and fun. Your goats will love you for it.[101]

Introducing New Goats to Your Herd

There are two main concerns when introducing a new goat to the herd:
1) The health of the goat. You don't want any parasites, infections, or diseases being introduced to your herd. The new goat also needs to transition from their old feed to your feeding program.
2) Your new goat's safety and wellbeing as they are integrated into the hierarchy.

The Health of the Goat: Quarantine to Protect Your Herd

You need time to get tests done and watch for signs of anything that might be contagious or cause the goat to present as "weak" when introduced to the herd. You want your goat to be at its best health and strength as they integrate into the goat hierarchy.

Make a space where the goat cannot touch or be touched by any member of the herd. (The other side of the fence is not adequate quarantine.) Make this space large enough to accommodate more than just the goats you are introducing so that after their quarantine, others can visit and get acquainted before turning the new goat out into the whole herd. (More details on that below.) If possible, set up the quarantine space where the new goat and the herd can see each other from a distance.

The recommended length of time for quarantine is 30-90 days. The length of time depends on the goat's tests (with vet papers to prove them).

Caprine Arthritis-Encephalitis Virus

Did the seller have papers to show that the goat recently tested negative for Caprine Arthritis-Encephalitis Virus (aka CAEV CAE)? If there is any uncertainty, get the goat tested before you purchase it or immediately after.

Coccidiosis

When introducing new goats to a herd, many goat keepers routinely go through a treatment cycle for coccidiosis for precaution, particularly if they are young.

Ask your vet about other de-worming treatments they might recommend before introducing your new goat to your herd. Many goat keepers include other routine de-worming treatments along with the coccidiosis treatment. The timing of these will be important, so consult with your vet and make a strategy.

The Health of Your New Goat: Transition the Feed

When you purchase your new goat, either buy a few weeks' feed from the owner or find out exactly what the goat has been eating so, you can replicate it. It's necessary to transition the goat to your usual feed for your goats rather than make a sudden change. A sudden change in diet can be hard on their digestive system and may result in symptoms such as diarrhea. These symptoms can be confused with actual illness or parasites. It can also stress the body of a goat who already has worms or is going through de-worming treatment.

Start with one or two days of the feed they are used to (the goat has had to adjust to a new place to live, that's enough for a couple of days). Then, gradually add in your feed over a couple of weeks until they are eating your feed exclusively. If the goat is healthy and the food seems to be disagreeable to them, then slow down the transition and adjust to the goat's responses. Usually, goats transition smoothly to the new feed; their stomachs just need a little time.

Introducing your Goat to the Herd [102]

When you know that your new goat is healthy, they are ready to meet the herd. Even though goats are intelligent, playful, delightful, funny, and adorable, when it comes to the herd, there is a hierarchy, and they can be brutal bullies.

Goats have been killed or beaten to the point of having to be put down when not introduced properly and left unsupervised. As prey animals who run in herds, the goat's primal instinct is to understand their place in the herd and remove any threat. That threat might be perceived weakness (that's one reason you want the new goat to be healthy and strong) or status (either a threat to current power or a target for not being assertive). There is a certain amount of just "working it out" that goes on that is normal, but "prevention is the best cure", and this doesn't have to be hard.

The good news is that goat owners regularly introduce new goats, and most goats adapt to a new herd well. Here we offer a couple of different ways to introduce new goats that are very effective in warding off dangerous situations and make the process simple and easy.

The best way to introduce your new goat to the herd is to introduce other goats individually and gradually. The goats you want to introduce are the ones in the middle of the hierarchy, not the ones at the top or necessarily the bottom. You can do this in several ways:

Option 1: Make your quarantine space large enough to accommodate 2-4 more goats – not for sleeping, but for a few hours a day while you are around to observe. Introduce one goat who is established as a "middle status" tier, then another; over the course of a few days to a week (depending on how things go), you may be able to allow in 2-4 of the herd in with the new goat.

The beauty of this is that your goat will have friends when they are let loose in the larger herd. Also, it is likely that those mid-level goats will now be protective of the new goat.

When the goat is ready to integrate into the paddock, contain the herd in a barn or another paddock if you can, and let the goat have a few hours with one of the new friends they've made to know where the water, food, baking soda, minerals, shelter and hang-out spaces all are. They can discover the boundaries of the fence and learn where to get their needs met without stress. Then let in the rest of the friends, then the rest of the herd.[103]

Option 2: If you don't have room to make extra space for more goats in the quarantine area, then after the quarantine, contain your herd in a barn or another paddock and let the goat into the space as described in the paragraph above with just one other goat.

The difference in this option is that you've not had the opportunity for the new goat to establish friends, so you need to do this process over a few days. Let one goat in who you think will be friendly who is in the middle of the hierarchy. Let them be for a few hours. The next day, add another one (or maybe two, depending on how it went). You want to build up gradually so that the new goat has a cluster of mid status friends before letting in the rest of the herd.

Expect to see some head butting and working things out. You don't need to step in unless you think your goat is in danger. Supervise, but don't intervene unless necessary.

When the new goat has a few friends, on the final day of introduction, let them into the paddock alone first if you have a place to enclose the other goats away from the paddock. The new goat can then sniff around, find out where the water, food, and shelter are. Next, let the friends inside. After an hour or so, let the rest of the herd into the paddock.

Again, supervise, and be ready to step in if necessary, but let the herd dynamics work themselves out. When you spend time with your herd, give your goats a lot of praise and attention so that the new goat is not considered "favored" and therefore resented. It really is a lot like being a new kid in school.

If the new goat hangs out on the edges, don't worry. As long as they can access water, food, and shelter, they can take their time making friends. They will observe and figure out how to integrate more in their own time.

Part 4: Breeding

You may have decided that you want to breed your goats. If you're going to have dairy without purchasing dairy goats regularly, you *have* to breed your does and dams to "freshen" them.

Breeding goats involves at least one buck. Bucks are, let's say, "a class in themselves," and you also need to be prepared to manage pregnancy, birth, and the care of baby goats before they are weaned.

This section will provide a firm fundamental knowledge base for breeding your goats and discuss how to troubleshoot before the trouble starts. With planning and attention, breeding your goats and raising kids with the genetics you want can be both practical and rewarding.

Seasons and Cycles of Goat Breeding

Not all goats breed in seasons. Some are seasonal, and some are year-round. One goat expert notes that the goats from the cold climates tend to breed seasonally from August-December, such as the LaMancha, Alpine, Oberhasli, and Saanen. The breeds from hotter climates often breed year-round, such as the meat goat breeds, Pygmys, and Nigerian Dwarfs.[104] Some Nubians will breed all year, and others will breed seasonally.

Even if your goats breed year-round, the length of daylight affects their reproduction. It is important to note that their most vigorous libido, highest fertility, volume, and quality of the semen peak in August-mid-September. [105]

Understand that even if you have goats that breed "seasonally", you can find yourself with surprise breeding, pregnancy, and surprise kids to be delivered. Don't think that because your goats are seasonal breeders, you don't need to keep the bucks and the does apart if you don't want kids.

Bucks

How many bucks should be in a herd?

Unless you have a large goat enterprise, you will need only one buck for your herd. Too many per doe capita can be a stress on both the bucks and the does. This will be clear when we look at how many does a buck can breed per season by age.

Choosing a Buck

When you choose a buck, do so very carefully. This one animal will have a massive influence on the genetics of your herd. In addition to health and any other attributes based on function (like working goats or fiber goats), temperament can be passed down. Have a thorough interview with your prospective buck. If you can, watch him with other goats and be observant of how he responds to you.

One important decision you will be making is whether you want to own a buck in order to breed. Owning has the advantages of being in control of their nutrition, health, and being able to watch them over time to make sure you want to breed this animal. Once he is out of quarantine when you've first acquired him, you don't have any concerns about introducing another unknown goat into your herd.

Renting a buck as a stud can be another option. This way, you don't have to have to live with a buck and manage his access to does when you don't want him to mate with them. The risk you have to assess is whether these are the genes you want in your herd and whether he has any communicable diseases or parasites that may be transferred to your doe (and, by extension, to the herd). You want to ensure that there are papers that prove that he tested negative for CL, *Caseous Lymphadenitis,* and CAE *Caprine Arthritis Encephalitis Virus.*

There is no test for chlamydiosis, and it can cause miscarriage in the last 2 months of pregnancy and stillbirths. If you want to breed your doe(s) with a billy goat outside your herd, determine whether the herd he comes from has had miscarriages or stillbirths. If he has been lent out to sire in other herds, it is not recommended that your does breed with this goat. Some professionals who rent out sires have rotations so that you can be assured by testing and the passage of time that this goat is an exception to this rule. Talk to your vet and experienced goat keepers in your area to find out about credible and trustworthy professionals who rent out bucks to sire.

If I rented a buck, I would want the testing verified for CL and CAE and written statements of when the goat was last de-wormed, that there had been no stillbirths or miscarriages in their herd and that the buck had not been rented out to another herd.

Some people make casual arrangements with other goat keepers in their area. Apply the same rigor of testing and de-worming verification in writing. There are even "driveway breedings" when you walk your goat in the driveway when the doe is in heat, and the buck is in rut, let them go at it, and then you're done. The central factor in these convenient, casual relationships is that you need to know the buck well enough to know that you want his genetics and that he is healthy. Asking for the formal "in writing" verifications and statements can seem awkward with a neighbor, which is why I have never done this. Consult with your vet. Get the factual information so you can properly assess your risk and make your decisions from there.

"Buck Power"

Buck power is a real phrase in the goat world. It is the number of does that a buck can breed during a season or month. This number varies by age, and it is essential to make sure that your young buck isn't overdoing himself.

Overbreeding a buck has two significant issues. One is that the volume and quality of the sperm go down with too much breeding. If the quality of sperm is not the best, then the kids' health can be sub-par. Another problem is that overbreeding can cause vulnerability and health issues for the buck. Breeding and rut will sap the goat's energy and bring on fatigue. The thing is, he won't know he is exhausted because he is in "all rut all the time" mode. He will neglect food, water, sleep, and rest and is obsessed and determined. Breeding and rut take a *lot* of energy from the buck, and even with the best nutrition and care, there's only so much breeding that is healthy in one time period.

Buck Power Breeding Numbers by Age

It is important to understand that bucklings can breed as early as 4 months.[106] Even though they *can*, does not mean that they *should*. The USDA cooperative extension[107] , as well as most goat experts and keepers, will say that they wait until they are 1 year old. The problem is, under a year, they are not fully developed, and their sperm will not be the best quality. If they breed with a doeling their size, she is likely to suffer stunted growth, miscarriage, injury (from not being ready to handle the weight of pregnancy), or even death.

Note that when a buckling is coming up on 3-1/2 months old, you need to prepare a space where he can be separated from the doelings. Watch him for signs of rut.

At age 1, a buck should only breed 10 does maximum in a month
At age 2, he can breed with 20-30 does.
At age 3+, he can breed up to 40 does
All of these numbers are dependent on proper nutrition and care.[108]

Conclusion of Bucks

If you want to breed your goats, then owning your buck will ensure that you are in control over the genetics and traits of the kids.

- Be prepared to manage the rut (it's so smelly!)
- Make sure you have an extra paddock to separate him when he needs to be away from the does (remember, they will breed through a fence, so the enclosure must not be adjacent to your does!)
- Don't breed him before 1 year and make sure he doesn't breed too much according to his age.
- Give him extra nutrition as discussed and observe him to make sure he is getting enough food and water.

Does

A doe coming into heat can make a buck go into rut and vice versa. A doeling can be catapulted into puberty too early for her health. Goats are fecund creatures; if you have dairy goats or does who are nursing, be aware that lactation does not stop them from getting pregnant. If the goat is nursing, this can be too soon; she should be allowed to rest rather than having her body energy go to both milk and pregnancy.

Age to Start Breeding Does

The gold standard for knowing when a doe is ready to breed is by weight. She can be bred when she is 60-70% of the average adult weight for her breed (for example, the average weight of a Nigerian Dwarf is 75 lbs., so a doe is ready to breed when she weights 45-52.5 lbs.) Sometimes does are ready around 8 months old.

Some goat keepers choose to wait until the second breeding season of the doe in order to allow them to fully grow, develop, and be at their full adult strength before kidding. You may have good reasons to get your doeling bred as soon as she is ready. If that is the case, then use the weight standard. If you can wait until she is at least a year, then you don't have to bother with weighing a goat. If you are willing to wait till her second season, all the better; she will be stronger.[109]

As mentioned above, it is critical to ensure that a doe is not bred before her body can safely carry the weight. Prior to her adequate development, pregnancy often results in stunted growth of the doeling, miscarriage, injury, or even death. [110]

Age to Stop Breeding Your Dams

Goats do not go through menopause. A dam *can* breed and get pregnant her whole life; they do not stop ovulating with age. The problem with breeding older dams is that they have a harder time in birth and are at a higher risk of death. The most humane thing to do is to isolate her from the bucks when she is 10. She will need to stay away from them for the rest of her life as she will never stop her estrus cycles.[111]

Tracking your Genetics

If you have a large enough herd, you may want to track which bucks have mated with which does. This can be important for pedigree or noting the characteristics that arise in the kids from particular matches.

You can keep a journal for this, but how can you keep track of which buck is the father of which kids? The answer: a mating harness, aka marker harness. These devices secure a crayon on the goat's chest so that he marks the doe on her back. This way, you can tell "who has been with who" when you were away from the barnyard and make a record of the mating. If the doe is pregnant, then you will know who the father is.

Here is one marking harness

Amazon "matingmark – Deluxe Breeding Harness"
http://ow.ly/V1Xy50CHgNJ

Here are the crayons

**Hot Purple, Hot Red & Hot Green
MATINGMARK Crayons PLUS
Red CEEMARK Marker**

Amazon – MatingMark Crayon Monitoring Breeding
http://ow.ly/Dzm950CHgVu

Choosing Breeding Partners

One common question is whether it is safe to allow goat siblings and families to breed – does it matter? The answer is yes, it does matter. Inbreeding can result in miscarriage or deformity. That is different from line breeding, which is a way to accentuate the genetic characteristics you want to bring out. Let's discuss the difference.

Inbreeding vs. Line Breeding

The difference between line breeding and inbreeding is that line breeding is breeding closely related goats in order to intensify the characteristics.

Inbreeding is when the bred pair is *too* close. The goat experts I've relied on have essentially said that they don't breed full brother and sister or father to daughter, but step brother to step sister is ok.

The consensus seems to be that it's line breeding if you get the desired results and inbreeding if you get unwanted results. Apparently, this distinction is not an exact science.

Line breeding can be a great way to intentionally intensify the characteristics that you want. You must also understand that it will intensify characteristics in those goats that you do **not** want. So choose carefully; it will intensify the characteristics across the board.[112]

How do I know when a doe is in heat, and a buck is in rut?

Chances are you won't need to read the answer to this question; it will be pretty obvious as you watch your goats.

Both does, and bucks are under the influence of powerful hormones when they are in heat or rut. They will alter their behavior to each other as well as exhibit personality changes.

Here's a list of behaviors you might see. All does exhibit some of them, but not necessarily all of them.

- Does often act restless and trot back and forth, especially to show off in front of the buck.
- Wagging her tail back and forth (this is called "flagging")
- Make more noise than usual
- Show more aggression, or mount other does
- Stands so a buck can breed with her (this may even be putting her backside against a fence)
- If you are milking your does, then the milk production will decrease during estrus.

Underneath, the doe's tail might be red or look swollen, or there may be a mucous discharge. This is normal if the doe is in heat.

The bucks are even more obvious. We've mentioned some of these behaviors earlier:
- Peeing on their legs, beard, and face
- Make noisy sounds (often raising the upper lip)
- Tongue Flapping
- Have a distinctive, strong repugnant odor
- Follow the does around and try to mount them
- They will eat less

Planning the timing

Timing may not matter to you, but if you have cold or wet winters, or if you are going to be away or unavailable to give attention to a birthing and newborn kids, then you will save yourself stress by timing your goat breeding.

Gestation for goats is roughly 5 months on average. [113] If you live in a place with cold or wet winters, then be aware that allowing your goats to breed in August will deliver you a kidding in January/February.

Nutrition during breeding

Breeding takes a lot of energy, and you want the embryos to be healthy and not vulnerable to conditions such as Nutritional Muscular Dystrophy that has been linked to selenium and vitamin E deficiencies. It is recommended that their diet is supplemented with vitamins and minerals. Talk to your vet about this, the requirements for does and bucks differ, and your usual routine for feeding, as well as the environment of your goats, will determine what kinds of changes will be right for you.

Pregnancy and Birth

It can be a little tricky to know whether a doe is pregnant. Sometimes you can tell by her shape, but some breeds (e.g., Nigerian Dwarfs) are "pot-bellied" anyway. I have a hard time looking at my Nigerian Dwarf does and am often wondering, "Is she pregnant?" just because I saw her from an angle that accentuated her usual roundness.

If your goats breed year-round and you track the female's cycles, then you may notice a cycle missed.

If you really want to know, you can talk to your vet about a blood, milk, or urine test.

Birthing supplies

There are a lot of lists of supplies for goat kidding on the internet; here are the ones that have served me best. Read others, figure out what works for you. [114]

*Please note: The following information about birthing is **not** veterinary advice; it is shared information from a fellow goat owner. You can use it as an introduction so that you are familiar with the process when you talk to your vet to get their advice. Look at these recommendations, compare others, and then talk to your vet about what they recommend.*

Note: There are items on the following list that require a vet's referral
- A clean, protected place for the doe to give birth (I like to have a bale of straw close to hand so that I can lay out a clean space for the new little ones.)
- A box of sterile nitrile gloves
- Hand sanitizer
- Fragrance-free baby wipes for hand cleaning
- Clean towels (at least one for each kid expected, more is better)
- Training pads for large dogs are great for laying down a newborn kid and absorbing some of the afterbirth fluids
- Sharp, sterile scissors for cutting the umbilical cord
- Dental floss for tying off the umbilical cord
- Ask your vet for a recommendation for applying to the cut end of the umbilical cord
- A surgical scrub for your hands and arms in the event that you need to reach inside the doe to help the birth. Ask your vet for a recommendation.
- Lubricant for livestock in case you need to reach inside the doe for the kid.
- Both headlamps and flashlights. They have different uses; both are handy. You should have them ready even if you have lighting, and sometimes you need to get light to a shadowed area.
- Suction bulb used for babies, sometimes the kid's nose or mouth needs clearing
- Thermometer – the easier to use, the better. Taking the kid's temperature will help with the health assessment.
- Bucket of warm water with soap
- Large heavy-duty lawn and leaf bags for garbage
- Bottles and teats in case you need them.
- Colostrum from a feed supply (it will require refrigeration) in case you need it.

How can I tell a doe is in labor?

4-6 weeks before kidding or as late as the last week.
The doe's udder will swell; goat keepers call this "bagging-up".

The kid "drops" (just like babies in women). In a goat's body, this is visible because her sides will sink in and become concave, and her hips will look bigger.

4 weeks – just before delivery – The tail ligaments begin to loosen. 24 hours before delivery, they are flaccid, and you can wrap your fingers around the tail head.[115]

- The tail has two ligaments, one on each side. This is an excellent sign of nearing labor, so get to know what a does tail feels like when she is not pregnant, then check when she's about 3.5 months and begin to tell the difference. This footnote has a .gif that shows you how to do it.[116]

Last 4 weeks: Her vulva begins to loosen and can look more prominent.

Last 4 weeks: Vaginal discharge is normal

Last 24 hours: The doe will seek solitude and want to find a nest. Try to have a space ready for her to be comfortable. She will also want to paw on the ground and move the bedding around. She's arranging things for the baby.

Last 24 hours: She will be getting up and down a lot – like she can't get comfortable.

Last 24 hours: Personality change. Some friendly does may become aloof, and some does who are standoffish typically may become friendly, affectionate, and seek physical contact for comfort. Give her the space to be who she is in that moment and know that this might be a sign that she's going to be kidding within a day.

Kidding (Birthing) [117]

The water bag may burst just before labor; individuals vary.

Stage 1 Labor: When the contractions begin, the doe will arch her back, and her tail will stick up.

Stage 2 Labor: The doe is pushing, and you'll see a kid begin to emerge.

The normal position is for the kid's feet to be face down. If you see a tail, then it is a breech.

Goats are strong and hardy creatures and often give birth without assistance. If an hour goes by and your doe is still pushing, you might want to be prepared to intervene. Ask your vet for instructions or resources to know what to do if the kid is breech or if the doe is not pushing the kid out.

It's not unusual for one kid to be born, then to have time between the next kid. It can be as long as 35 minutes before the next contractions begin.

Before your doe gives birth, talk to your vet about all the "what ifs". If you know have a local mentor who is experienced with kidding, they are also a good resource and may even be willing to be on call to help if you have problems.

Afterbirth Care

Umbilical Cord

Trim the kid's umbilical cord to no shorter than 2". You don't want it to be left long as it is a place of entrance for bacteria, and the longer it is, the more chance there is that it will get into feces, etc. Tie it off with the floss.

Warm and Dry

As tough as goats are generally, they are very fragile as newborns. Hypothermia is a high risk. Keep your newborn kids warm and dry. Some goat keepers use a heat lamp for their newborns in this vulnerable stage.

Getting Milk from Mama

After birthing, the doe's milk right after birth is colostrum. The kid must get this as soon as they can after birth because the colostrum has both vital nutrition as well as components that build immunity, which is foundational for their health and strength.

Often, this will happen easily and naturally. Too much intervention will interrupt their important bonding process. Let them be, watch to ensure that the mother is showing care by licking, etc. The kid usually starts sucking, and all is good. Let nature take its course, there is a learning curve for the kid, but that's normal.

Watch for any sign that the teat might be blocked. If it is, try gently milking it. Use a warm damp cloth on it to help get the blockage to warm up and get out if the milking doesn't work.

If an hour is coming up and the kid is not feeding, then help them out. If it's not working, then milk the colostrum from the doe and bottle feed it to the kid.

Ask your vet about the recommended schedule for vaccinations, disbudding (needs to be in the first 10 days, some goat owners recommend no longer than 3), or castration (if you choose to castrate your bucks).

This footnote includes curated resources for post-partum care. One of them also includes a very useful timeline for the baby goats. [118]

Mother rejecting kid

It does not happen too often, but occasionally a mother goat will reject their kid. Sometimes it is one kid born alone; other times, the mother will nurture one and reject the other(s).

When kid rejection happens, it can be shocking and heartbreaking. Dams will sometimes butt and bite on an ear of the kid to literally throw them. I've seen forums where the goat owner will say they were afraid to leave the kid overnight with the mother for fear of the mother killing it.

We can speculate on the reasons that a dam might reject a kid. Many mammals have an instinct to reject a diseased offspring that they know or expect will die. It may be the mother's discomfort on her teats. Some goat owners suggest that particularly difficult labors may be the reason for rejection.

Whatever the reason, you need to be prepared for the "what if". You need to be prepared to bottle feed (and lose some sleep) and to work with the kid and dam to see if she can be persuaded.

If a dam rejects a kid at birth, make sure that the kid gets the colostrum and is warm and dry. Then give it some time and observe. Make sure that the udder and teats of the dam are not painful to touch or to milk. If they are, she may have mastitis or a blockage. She will certainly kick the kid off if sucking hurts her.

Take the temperature of both the dam and the kid. A normal temperature is 101.5-103.5F. If it is above that range, there may be an infection; if it is below 100F, the kid is in danger of hypothermia. Any perceived weakness in the kid can sometimes trigger a rejection response in a dam.

Remove the kid from the dam if she is being aggressive and may harm or kill the kid.

Supplement with milk, so the kid gets some nutrition.

Sometimes everything will go well for days; the kid is happily sucking, all is maternal bliss, and then suddenly the dam gets aggressive and will reject the kid. If this happens, it is more likely to do with mastitis or discomfort of the dam, but it is worth checking on the kid as well. Take the dam's temperature; if it is raised, it may indicate mastitis or another infection causing pain or discomfort. Also, check her udder and teats.

As well as checking on the dam, take the kid's temperature and check the mouth. Sometimes they can have a loose tooth that is sharp on the dam's teat.

Sometimes the problem can be the sucking technique of the kid. If the kid is not latching on to a good mouthful of the teat and is sucking on the end of the nipple instead, it will be very painful for the dam. Here is an article about kids and sucking techniques as well as photos of what to look for.[119]

Most goat owners say that attention, patience, and time are required to help-out in this sad situation. If you feel it is safe, then enclose the dam and the kid together in a small, warm, clean shelter.

Horns or Disbudding – The Pros and Cons

If you breed your goats, then you will have to decide whether or not to disbud. Disbudding is one of those areas where you will find strong opinions on both sides. It seems to come down to the context and preferences of the goat owner.

- **Pros of Horns**
  - The goat has a way to defend itself from predators.
  - Horns are a great "handle" to grab onto. (Some goat owners say they won't have goats without them for this reason.)
  - Horns have a lot of blood vessels in them and help the goat to regulate their temperature. Without horns, the goat will need to pant to cool off.[120]
  - Horns are beautiful.

- **Cons of Horns**
  - Horns can get caught in fences, feeders, and places you wouldn't think of. You must be proactive about this to make sure your fences and equipment are safe for the goats.
  - Horns can be an unfair advantage if you have others in the herd who do not have horns.
  - Horns can be used to hurt you if the goat is not well trained.

- **Pros of Disbudding**
  - Removes the danger of the horns to others in the herd as well as yourself.
  - There are no horns to get stuck in the fence, the feeder, and other places.

- **Cons of Disbudding**
  - You don't have those handy handles to grab onto.
  - The goat cannot defend itself against predators or other horned members of the herd.
  - It's another thing to do in that busy first week after a kid is born. If you have multiple kids multiple times a year, that's extra work.
  - If you live in a hot climate, you've lost that temperature regulation. More care needs to be taken to ensure that they do not overheat.

You can see that some of the criteria for whether or not to disbud may have to do with environment and function. For example, the brush goat professionals I've met have all their goats disbudded because of the danger of their goats getting caught in the electric fence or in the brush itself. They use guardian livestock dogs for predator protection. If you plan on showing your goats, you will likely be required to disbud them.

On the other hand, my neighbors breed Nigerian Dwarfs and have decided to let their goats have horns. They breed very carefully for temperament and have never had the horns be anything but a plus for the purposes of predation and as a handle.

Most pack goat owners want their goats to be horned so they can defend themselves in the backcountry if necessary and have the help of the temperature regulation when they are exerting themselves.

If you decide to let your goats have their horns, then consider looking for horns on any goat you incorporate into the herd. Introducing a new goat with no horns to an entire horned herd is unfair and could cause serious problems for the new goat.

If you want to disbud, you *can* do it yourself, but don't try to learn through YouTube videos. Ask your vet or an experienced goat breeder to walk you through it at least once. The procedure is performed with a disbudding iron that burns off the budding horns.[121]

Part 5: Conclusion

Goats are delightful and relatable creatures. Most breeds are tough and hardy. They don't consume the water or food that larger livestock do and have the intelligence to be trained. It is possible to breed goats for good temperaments as well as other characteristics you may want.

Most of the common health conditions that goats suffer are preventable or treatable. Something like a breech birth can be managed safely if you are prepared.

Farm animals are always a responsibility and a risk. Goats are no exception. Nothing with the word "farm" involved is ever "stress-free", but you can design a lot of ease into your goat keeping and increase your joy If you follow the preventative guidelines in this book.

As discussed, make sure you have a good vet and, if possible, experienced local goat keepers to talk to. The internet is full of great information as well as contradictory and confusing information. Videos can help you get a sense of learning about activities like kidding but use them to be informed before you speak to someone you trust in person. There is nothing like true mentorship and experience.

Whether it be as a pet for affection and connection, for milk, for meat, for show, for agility, for fiber, for brush-clearing, or for pulling carts and packing in the backcountry, enjoy the rewards and all the hilarity that comes with it.

Endnotes

[1] https://goats.extension.org/chlamydiosis/

https://www.vet.cornell.edu/animal-health-diagnostic-center/programs/nyschap/modules-documents/zoonotic-diseases-sheepgoats

https://abga.org/wp-content/uploads/2016/01/Chlamydia.pdf

https://www.goatfarming.ooo/2018/07/abortion-in-goats-with-focus-on.html

[2] Ibid.

[3] https://www.mayoclinic.org/diseases-conditions/q-fever/symptoms-causes/syc-20352995

[4]https://www.mayoclinic.org/diseases-conditions/q-fever/symptoms-causes/syc-20352995

https://healthtian.com/q-fever/

https://www.healthline.com/health/q-fever

[5] https://www.cfsph.iastate.edu/FastFacts/pdfs/leptospirosis_F.pdf

[6] https://www.goatgyan.com/33-news/212-leptospirosis-in-goats

[7] https://goats.extension.org/ringworm/

[8] Salmonella resources:
https://www.extension.purdue.edu/extmedia/AS/AS-595-commonDiseases.pdf

http://goat-link.com/content/view/185/183/

http://www.goatworld.com/articles/health/commondiseases.shtml

[9]
https://www.cdc.gov/salmonella/general/index.html#:~:text=Most%20people%20with%20Salmonella%20infection,experience%20symptoms%20for%20several%20weeks.

[10] https://goats.extension.org/goat-reproduction-puberty-and-sexual-maturity/

[11] Ibid.

[12] https://simplelivingcountrygal.com/how-to-keep-a-buck-so-you-can-breed-your-goats/

[13] By "common" I mean typically accessible for purchase in the U.S.

[14] https://modernfarmer.com/2014/06/pygmy-goats-were-made-to-prance/

[15] https://www.roysfarm.com/african-pygmy-goat/

https://goats.extension.org/goat-breeds-pygmy/

https://rurallivingtoday.com/livestock/everything-you-need-to-know-about-pygmy-goats/

[16] https://animalsake.com/pygmy-goat-care

[17] https://farminence.com/the-complete-guide-to-pygmy-goats/#Do_you_own_pygmy_goats_How_do_you_take_care_of_them_How_are_they_different_from_standard_sized_goats

[18] https://www.newlifeonahomestead.com/nigerian-dwarf-goats/#Nigerian_Goat_History

[19] Ibid.

[20] https://www.wideopenpets.com/all-you-need-to-know-about-the-nigerian-dwarf-goat/

[21] https://www.wideopenpets.com/all-you-need-to-know-about-the-nigerian-dwarf-goat/

[22] https://www.pba-pygora.org/

[23] https://www.livescience.com/34435-fainting-goats.html

https://www.boergoatprofitsguide.com/raising-fainting-goats/

[24] https://www.hobbyfarms.com/myotonic-goats-3/

[25] https://en.wikipedia.org/wiki/Fainting_goat

[26] https://sweetgumminifarm.webstarts.com/minis.html

[27] https://morningchores.com/american-lamancha-goats/

[28] https://www.goatfarmers.com/blog/brush-goats-brush-control-land-clearing%2F

[29] http://operationhomestead.blogspot.com/2011/07/jurassic-goat.html

[30] http://operationhomestead.blogspot.com/2011/07/jurassic-goat.html

[31] https://www.youtube.com/watch?v=8NG6BWhb6rs

[32] Resources for plants that are toxic to goats

https://poisonousplants.ansci.cornell.edu/goatlist.html

https://content.ces.ncsu.edu/poisonous-plants-to-livestock

http://goatworld.com/health/plants/hemlock.shtml

https://packgoats.com/toxic-plants-for-goats/

https://backyardgoats.iamcountryside.com/feed-housing/poisonous-plants-for-goats-avoiding-dastardly-disasters/

[33] https://www.youtube.com/watch?v=qdGB_vF6UAY

[34] https://bestreviews.com/best-electric-fence-chargers

https://ecotality.com/best-solar-electric-fence-chargers/

https://cleanenergysummit.org/best-solar-fence-chargers/

https://greencoast.org/best-solar-fence-chargers/

[35] https://www.goatfarmers.com/blog/brush-goats-brush-control-land-clearing%2F

[36] https://bootsandhooveshomestead.com/alpine-goats/
https://www.boergoatprofitsguide.com/raising-alpine-goats/

[37] See the section on training your goat for avoiding and handling stubbornness.

[38] Saanen Goats:
http://afs.okstate.edu/breeds/goats/saanen/
https://www.roysfarm.com/saanen-goat/
https://morningchores.com/saanen-goat/

[39] https://goats.extension.org/goat-breeds-nubian/

[40] Metal Stanchions: https://www.valleyvet.com/ct_detail.html?pgguid=da29ad59-e97f-470b-85a0-a238be334467&itemguid=8f2785e9-509c-4864-b81d-dcd2879b2bd6&sfb=1&grp=3000&grpc=3400&grpsc=3430&sp=f&utm_content=38608&ccd=&utm_source=&utm_medium=&msclkid=0772945a418d133d84edcbd1267cad9a&utm_campaign=F%20Cat%20Goat%20Sheep%20(3000)%20Equipment%20(3400)%20v2%20PLA&utm_term=4584619891003831
https://www.premier1supplies.com/p/milking-stand?msclkid=8ae753f160c21bb319600c7a07331cbb&utm_source=bing&utm_medium=cpc&utm_campaign=(ROI)%20Shopping%20-%20Sheep%20%26%20Goats&utm_term=4584001426103569&utm_content=Sheep%20%26%20Goats

A video tutorial showing you how to make your own milking stanchion
https://www.youtube.com/watch?v=gqBYj0ipjwU

Wood stanchion that you assemble
https://www.amazon.com/Goatstandcom-Large-48x22-Carpenter-Stanchion/dp/B01KEDHT10

More instructions for building your own
https://packgoats.com/courses/goat-milking-and-trimming-stanchion-plans/

[41] https://www.youtube.com/watch?v=tnA6YqT_7aM
https://www.youtube.com/watch?v=RWswEEbMIL4

[42] Ibid.

[43] https://www.bewellbuzz.com/body-buzz/nutrition/11-health-benefits-of-raw-goat-milk/

[44] https://www.fda.gov/food/buy-store-serve-safe-food/dangers-raw-milk-unpasteurized-milk-can-pose-serious-health-risk

[45] https://cityfarmingbook.com/milking-dairy-goats/

[46] https://www.youtube.com/watch?v=kODA_hOVW9M

[47] https://www.ithacaweek-ic.com/ithaca-fiber-farm-grows-with-help-from-hybrid-cashgora-goats/
https://www.goatfarmers.com/blog/fiber-goats-breeding-production-sales%2F

[48] https://morningchores.com/cashmere-goat/

[49] Ibid.

[50] Ibid.

[51] Ibid.

[52] https://homestead.motherearthnews.com/making-money-angora-goats-part-4/

[53] https://www.lifestyleblock.co.nz/lifestyle-file/livestock-a-pets/goats/angora-goats/item/908-angora-health-diseases-of-the-skin-and-the-brain

[54] "*Ill-thrift* is a term used to describe when stock grow at a slower growth rate than expected, given their feed allocation. In this project it is defined as when lambs or young cattle have more than 30% slower growth rates than expected." https://beeflambnz.com/sites/default/files/factsheets/pdfs/Fact-Sheet-177%E2%80%94Ill-thrift.pdf

An article on general Angora health issues:
https://www.lifestyleblock.co.nz/lifestyle-file/livestock-a-pets/goats/angora-goats/item/909-angora-health-ill-thrift-and-miscellaneous-diseases

[55] https://www.lifestyleblock.co.nz/lifestyle-file/livestock-a-pets/goats/angora-goats/item/866-angora-health-lameness-and-sudden-death

[56] https://www.lifestyleblock.co.nz/lifestyle-file/livestock-a-pets/goats/angora-goats/item/803-angora-goats-health-and-disease-scouring
[57] https://morningchores.com/angora-goat/

[58] http://www.ladocumentationcaprine.net/plan/socio/art/page0631.pdf
https://www.mla.com.au/globalassets/mla-corporate/generic/extension-training-and-tools/gig-mohair.pdf

[59]Cashgora farmers
https://www.ithacaweek-ic.com/ithaca-fiber-farm-grows-with-help-from-hybrid-cashgora-goats/
https://mokafarm.com/meet-the-goats/cashgora/

A discussion on a forum regarding cashgora. See the responses from both "sweetgoats" and "keren". There is good information from experienced goat raisers here.
https://www.thegoatspot.net/threads/australian-cashmere-vs-cashgora.104558/

[60] Fiber Goat Associations

American Angora Goat Breeders Association (AAGBA)
http://www.aagba.org/

American Colored Angora Goat Registry
https://www.acagr.us/

American Nigora Goat Breeders Association
https://nigoragoats.homestead.com/

Cashmere Goat Association
https://cashmeregoatassociation.org/

Colored Angora Goat Breeders Association (CAGBA)
https://www.cagba.org/

Eastern Angora Goat & Mohair Association (EAGMA)
http://angoragoats.com/

Northwest Cashmere Association
http://www.nwcashmere.org/

Pygora Breeders Association (PBA)
https://pba-pygora.org/
American Goat Federation
https://americangoatfederation.org/

Cashmere and Camel Hair Manufacturers Institute (CCMI)
https://www.cashmere.org/about-us.php

Australian Cashmere Growers Association
https://www.australiancashmere.com.au/

The Angora Goat Society (U.K.)
www.angoragoats-mohair.org.uk

[61] This resource includes a discussion about whether or not to use a bit
https://backyardgoats.iamcountryside.com/kids-corner/training-goats-to-pull-carts/

This resource recommends adding a step before attaching the goat to the wagon – practicing with a "travois". If your goat is particularly skittish, this may help.

https://www.dummies.com/home-garden/hobby-farming/raising-goats/how-to-train-a-goat-to-pull-a-cart/

This woman is experienced and explains her process as well as her home-made cart
https://thriftyhomesteader.com/working-goats-your-journey-begins-here/

[62] I learned this the hard way when I had a goat who took to the collar and the bridle easily, sniffed the harness and seemed relaxed and nonchalant, but bucked, gave me a sudden swift kick and ran away when I introduced the harness. Tying the goat to the fence means that you can reassure them and determine whether to continue or how to change the harness introduction.

[63] The information in the pack goat section is (like all other sections) a mix of my own experience, the teaching from mentors and veterinarians, and credible articles online written by goat experts.

If you are interested in pack goats, the absolute best online source is Mark Warnke at http://www.packgoats.com His information is consistent with what I've learned from others. He has deep experience and a lot of wisdom. If you want to consider pack goats, go to his site and check out the courses he offers.

[64] https://www.highuintapackgoats.com/about-breeds.htm

[65] https://packgoats.com/goats-for-sale/

[66] https://www.hairstoncreekfarm.com/how-long-do-goats-live/

[67] https://packgoats.com/pack-goat-conditioning-and-weight-bearing-timeline/

[68]Ibid.

[69]Ibid.

[70] https://packgoats.com/training-baby-pack-goats/
https://packgoats.com/pack-goat-conditioning-and-weight-bearing-timeline/

[71] https://packgoats.com/pack-goat-conditioning-and-weight-bearing-timeline/

[72] https://packgoats.com/product/pack-goat-training-saddle-by-marc-warnke/

[73] https://packgoats.com/pack-goat-conditioning-and-weight-bearing-timeline/

[74] Goat Showing Tips – this is simple and a great introduction

https://www.dummies.com/home-garden/hobby-farming/raising-goats/ten-tips-for-showing-goats/This article goes into more detail about care and preparation necessary for showing a goat.
https://www.roysfarm.com/how-to-show-a-goat/

The American Goat Society offers information about scoring as well as a figure of the patterns expected when you walk your goat for the judges.
https://americangoatsociety.com/showmanship.php

Series on Market Goat Showmanship published by Washington State University written by a 4-H program coordinator.
Part I https://s3.wp.wsu.edu/uploads/sites/2075/2013/03/Market-Goat-Showmanship-Part-I.pdf

Part II https://s3.wp.wsu.edu/uploads/sites/2075/2013/03/Market-Goat-Showmanship-Part-II-web.pdf

Part III https://s3.wp.wsu.edu/uploads/sites/2075/2013/03/Market-Goat-Showmanship-Part-III-web.pdf

Note: even though these articles are about "market goats", many of the principles apply to any kind of goat showing. It's well worth the read.

An expert provides advice on prepping for show season
https://farmtek.wordpress.com/2013/06/11/its-show-time-tips-and-tricks-for-preparing-your-goats-for-a-successful-show-season/

Here is very helpful detailed advice about showing wethers.
http://showwethers.tripod.com/showmanship.html
Again, even though this is written specifically about wethers, there are many pointers here for anyone showing a goat.

At the bottom of the link about wethers above is more information about showing meat goats.

[75] https://morningchores.com/kiko-goat/
American Kiko Goat Association
https://kikogoats.com/

[76] https://content.ces.ncsu.edu/nutritional-feeding-management-of-meat-goats

77 https://www.goatfarmers.com/blog/feeding-goats-guide-nutritional%2F

.

[78] https://www.boergoatprofitsguide.com/goat-fencing/

[79] https://www.boergoatshome.com/hooves.php

[80] As you might expect, the "Goat Coat Shop" people are experts for goat jackets. They offer optional fleece linings so you can mix and match your outer water protection with warmth as you need.
http://www.goatcoatshop.com/

Valley Vet is also a reputable company and offers a number of goat jackets.
https://www.valleyvet.com/ct_detail.html?pgguid=09fb22d0-101f-4742-8b88-
e8781cb2dcb0&itemguid=9fbc8f92-7104-489b-8a6e-
c9fe657a974a&sfb=1&grp=L000&grpc=L700&grpsc=L720&sp=f&utm_content=45248&ccd=&utm_source
=&utm_medium=&msclkid=2a38def3f9a81783b255815a68661062&utm_campaign=F%20Cat%20Livestoc
k%20Grooming%20(L000)%20Show%20Goat%20(L700)%20v2%20PLA&utm_term=4585581966414345

[81] http://www.boergoats.com/clean/articleads.php?art=490

[82] https://morningchores.com/get-your-goats-ready-for-winter/

[83] https://goats.extension.org/caprine-arthritis-encephalitis-virus-cae/

[84] Ibid.

[85] https://www.canr.msu.edu/news/coccidiosis_in_goats_and_sheep

[86] Read more about wether stone formation here:
https://www.betterhensandgardens.com/goat-wethers-urinary-calculi/
https://hoeggerfarmyard.com/urinary-calculi-in-goats/

[87] https://www.amazon.com/NAYLOR-CO-INC-032584945156-
Antibacterial/dp/B001CD1HVK/ref=sr_1_2?dchild=1&hvadid=77790509631693&hvbmt=be&hvdev=c&hv
qmt=e&keywords=hoof+n+heal&qid=1608501031&sr=8-2&tag=mh0b-20

[88] https://goats.extension.org/goat-bloat/

[89] Ibid.

[90] Here are 2 University articles on external parasites that infest goats:
https://goats.extension.org/goat-external-parasites-arthropods/
https://extension.okstate.edu/fact-sheets/external-parasites-of-goats.html

These articles are by experienced and credible goat experts regarding external parasites:
https://www.betterhensandgardens.com/identify-control-goat-external-parasites/
https://morningchores.com/goat-external-parasites/

[91] The following is a list of links to articles regarding Caseous Lymphadenitis:

https://u.osu.edu/sheep/2019/01/15/caseous-lymphadenitis-cl-in-sheep-and-goats/
https://goats.extension.org/caseous-lymphadenitis/
https://www.boergoatprofitsguide.com/cl-in-goats/
https://www.farmhealthonline.com/US/disease-management/goat-diseases/cl-in-goats/

[92] https://goats.extension.org/caseous-lymphadenitis/

[93] Pneumonia:
https://www.boergoatprofitsguide.com/pneumonia-in-goats/
https://tennesseemeatgoats.com/articles2/pneumonia06.html

[94] Another set of clear instructions for introducing your dog to goats
https://dogcare.dailypuppy.com/introduce-dog-goats-3423.html

Excellent article on the general dynamics at play between dogs and goats
https://www.hobbyfarms.com/will-your-dog-attack-livestock-3/

Videos:
Scroll down to the second video on this page for how to introduce your dog to goats.
https://thedogvisitor.com/qa/how-do-you-introduce-a-dog-to-a-goat

[95] Livestock Guardian Dog Resources

An excellent introductory article that includes points about their behavior and how to decide whether an LGD is for you.
https://kingskeepfarm.com/livestock-guardian-dogs-what-you-need-to-know/

A great article by an experienced LGD breeder
https://permies.com/t/129855/Livestock-Guardian-Dogs-Small-Farmsteads

Breeds to Consider:
https://farmhouseguide.com/do-goats-and-dogs-get-along/

https://morningchores.com/farm-dogs/
https://www.thesprucepets.com/choosing-a-livestock-guardian-dog-breed-3016777
https://www.101dogbreeds.com/working/livestock-guardian-dogs

[96] https://coloradomountaindogs.com/

[97] https://kingskeepfarm.com/livestock-guardian-dogs-what-you-need-to-know/

[98] These people train Livestock Guardian Dogs and can assist you remotely.
https://dogfolksdirtfarm.com

[99] https://packgoats.com/training-your-pack-goat-kid-everything-your-pack-goat-will-need-to-learn-year-one/

[100] https://packgoats.com/how-to-train-a-goat-with-a-squirt-bottle/

[101] Resources for goat training
These are orientated towards pack goats, but he writes a lot about general principles for training goats and their psychology.

https://packgoats.com/training-your-pack-goat-kid-everything-your-pack-goat-will-need-to-learn-year-one/

https://packgoats.com/pack-goat-training/

https://packgoats.com/training-baby-pack-goats/

How to Train a Goat with a Squirt Bottle (includes video)
https://packgoats.com/how-to-train-a-goat-with-a-squirt-bottle/

How to Train your Goats to use a Litter Box (!)
https://morningchores.com/goat-litter-box-training/

[102] Here are some well written articles about introducing a new goat to your herd. They offer a bit more detail and are resonant with the advice I've received and practices I've incorporated from experienced goat keepers.

There is more detail about how to introduce does, vs bucks vs wethers in this article
https://morningchores.com/introduce-a-new-goat-to-an-existing-herd/

This short article includes a tragic story that demonstrates why it is important to introduce new goats with care and supervision
https://www.tractorsupply.com/tsc/cms/life-out-here/the-barn/animal-medication-for-goats/introducing-a-new-goat-into-a-herd

This article suggests bringing all the goats to neutral turf for introductions.
https://hoeggerfarmyard.com/introducing-new-goats-herd/

[103] Ibid.

[104] https://www.goatfarmers.com/blog/goat-breeding%2F
[105] https://goats.extension.org/goat-reproduction-puberty-and-sexual-maturity/

[106] https://goats.extension.org/goat-reproduction-puberty-and-sexual-maturity/

[107] Ibid.

[108] Ibid.

[109] https://www.goatfarmers.com/blog/goat-breeding%2F

[110] https://goats.extension.org/goat-reproduction-puberty-and-sexual-maturity/

[111] Ibid.
https://www.goatfarmers.com/blog/goat-breeding%2F

[112] https://www.weedemandreap.com/goat-breeding-101/

https://fiascofarm.com/goats/breeding.htm#Linebreeding
https://onpasture.com/2014/10/20/breeding-matters-iii-inbreeding-vs-line-breeding/

[113] https://goats.extension.org/goat-reproduction-puberty-and-sexual-maturity/

[114] https://www.boergoatprofitsguide.com/goat-supplies/
https://www.betterhensandgardens.com/kidding-supplies-and-preparation/
https://www.weedemandreap.com/goat-pregnancy-delivery-checklist/

[115] https://alifeofheritage.com/farm-living/goat-labor-signs/

[116] https://alifeofheritage.com/farm-living/goat-labor-signs/

[117] This is a resource on kidding with gifs and clear instructions.
https://alifeofheritage.com/farm-living/goat-labor-signs/

[118] This article includes great advice for post-partum care as well as a helpful timeline for baby goats.
https://www.weedemandreap.com/post-partum-care-goats/

Other excellent post-partum and baby goat care resources:
https://morningchores.com/baby-goat-care/
https://www.wikihow.com/Care-for-Baby-Goats
https://theorganicgoatlady.com/how-to-care-for-your-newborn-goat/

[119] https://antiquityoaks.blogspot.com/2008/06/sucking-disorder-in-goat-kid.html
[120] https://104homestead.com/pros-cons-disbudding-goat/

[121] https://www.weedemandreap.com/how-disbud-dehorn-baby-goat/

Printed in Great Britain
by Amazon

74916139R00088